Lecture Notes in Computer Science   5392

Commenced Publication in 1973
Founding and Former Series Editors:
Gerhard Goos, Juris Hartmanis, and Jan van Leeuwen

T0223384

David Duke   Lynda Hardman
Alex Hauptmann   Dietrich Paulus
Steffen Staab (Eds.)

Semantic
Multimedia

Third International Conference on Semantic
and Digital Media Technologies, SAMT 2008
Koblenz, Germany, December 3-5, 2008
Proceedings

 Springer

Volume Editors

David Duke
School of Computing, University of Leeds
Leeds, United Kingdom
E-mail: djd@comp.leeds.ac.uk

Lynda Hardman
CWI, Science Park Amsterdam
Amsterdam, The Netherlands
E-mail: lynda.hardman@cwi.nl

Alex Hauptmann
School of Computer Science, Carnegie Mellon University
Pittsburgh, PA, USA
E-mail: alex@cs.cmu.edu

Dietrich Paulus
Fachbereich Informatik, Universität Koblenz-Landau
Koblenz, Germany
E-mail: paulus@uni-koblenz.de

Steffen Staab
Fachbereich Informatik, Universität Koblenz-Landau
Koblenz, Germany
E-mail: staab@uni-koblenz.de

Library of Congress Control Number: Applied for

CR Subject Classification (1998): H.5.1, H.4, H.5, I.2.10, I.4, I.7, C.2

LNCS Sublibrary: SL 3 – Information Systems and Application, incl. Internet/Web and HCI

ISSN   0302-9743
ISBN-10   3-540-92234-2 Springer Berlin Heidelberg New York
ISBN-13   978-3-540-92234-6 Springer Berlin Heidelberg New York

Springer is a part of Springer Science+Business Media

springer.com

© Springer-Verlag Berlin Heidelberg 2008
Printed in Germany

Typesetting: Camera-ready by author, data conversion by Scientific Publishing Services, Chennai, India
Printed on acid-free paper   SPIN: 12591855   06/3180   5 4 3 2 1 0

Preface

We are pleased to welcome you to the proceedings of the Third International Conference on Semantic and Digital Media Technologies held in Koblenz, Germany. The SAMT agenda brings together researchers at extreme ends of the semantic multimedia spectrum. At one end, the Semantic Web and its supporting technologies are becoming established in both the open data environment and within specialist domains, such as corporate intranet search, e-Science (particularly life sciences), and cultural heritage. To facilitate the world-wide *sharing* of media, W3C is developing standard ways of denoting fragments of audio/visual content and of specifying and associating semantics with these. At the other end of the spectrum, media analysis tools continue to grow in sophistication, identifying features that can then be associated with explicit semantics, be they expressed formally or informally, using proprietary formats or open standards. Recent progress at these two fronts of the SAMT spectrum means that research spanning the semantic gap is now of vital importance to feed the real applications that are emerging.

This conference also represents a step towards bridging the gap between the research cultures and their respective approaches at both ends of the spectrum. The papers selected show that SAMT is able to attract researchers from media analysis, who see the benefits that more explicit semantics can provide, as well as researchers from knowledge engineering who realize that, while a picture can be expressed as a thousand concepts, a million more are waiting to be extracted. In order to attract high-quality papers at future conferences it is important that SAMT be recognized as the leading conference on semantic multimedia. To achieve this goal, this year's conference brought together the very best of the work in the area. Of the 52 full papers submitted to the conference, 12 were selected for inclusion in these proceedings.

Apart from the accepted full papers, the SAMT program was enriched by 34 poster presentations, selected from 46 submissions, a special demo session with 4 project demos, 2 keynote talks by Jane Hunter and Mor Namaan, as well as 2 invited talks given by EC representatives, Albert Gauthier and Luis Rodríguez-Roselló. Moreover, three interesting workshops were offered, namely:

- International Workshop on Interacting with Multimedia Content in the Social Semantic Web
- Cross-Media Information Analysis, Extraction and Management
- Semantic 3D Media

In addition, there were three tutorials, namely:

- A Semantic Multimedia Web
- Ontologies and the Semantics of Visualization
- Creating Semantically Rich Virtual Reality Applications

We would like to express our gratitude to everyone who contributed to the technical and organizational success of the SAMT 2008 conference. Special thanks go to the Poster and Demo Chairs, Michela Spagnuolo and Siegfried Handschuh, the Workshop and Tutorial Chairs, Fabio Ciravegna and David Duce, the Publicity Chair, Marcin Grzegorzek, and the Registration Chair, Ruth Götten, who all did a great job in the preparation phase for this event. Furthermore, we would like to thank the Program Committee members for a smooth review process, the invited speakers and tutors, the Industry Day Co-chair, Werner Haas, as well as all contributors and participants of the conference. Moreover, we are grateful for the generous contribution of our sponsors, namely, the consortia of the K-Space and WeKnowIt projects, the University of Koblenz-Landau, as well as the Ministry of Education, Science, Youth, and Culture of the German federal state of Rhineland-Palatinate.

This year's conference provided a high-quality archival publication and a varied and interesting program, designed to stimulate interaction among participants. For future conferences the challenge will be to encourage more researchers to contribute to the field, for example, researchers from 2&3D graphics, sensor technology, and further application domains.

December 2008

David Duke
Lynda Hardman
Alex Hauptmann
Dietrich Paulus
Steffen Staab

Organization

SAMT Program Committee Members

Yannis Avrithis, Greece
Bruno Bachimont, France
Mauro Barbieri, The Netherlands
Stefano Bocconi, Italy
Susanne Boll, Germany
Nozha Boujemaa, France
Oscar Celma, Spain
Lekha Chaisorn, Singapore
Philipp Cimiano, Germany
Michael Christel, Germany
Matt Cooper, USA
Thierry Declerck, Germany
Franciska de Jong, The Netherlands
Arjen DeVries, The Netherlands
Pinar Duygulu-Sahin, Turkey
Wolfgang Effelsberg, Germany
Joost Geurts, France
William I. Grosky, USA
Werner Haas, Austria
Allan Hanbury, Austria
Alan Hanjalic, The Netherlands
Michael Hausenblas, Austria
Paola Hobson, UK
Winston Hsu, Taiwan
Ichiro Ide, Japan
Antoine Isaac, The Netherlands
Ebroul Izquierdo, UK
Alejandro Jaimes, Spain
Joemon Jose, UK
Mohan Kankanhalli, Singapore
Brigitte Kerherv, Canada
Stefanos Kollias, Greece
Yiannis Kompatsiaris, Greece
Wessel Kraaij, The Netherlands
Hyowon Lee, Ireland
Jean Claude Leon, France
Paul Lewis, UK
Craig Lindley, Sweden

Suzanne Little, Germany
Vincenzo Lombardo, Italy
Nadja Magnenat-Thalmann,
  Switzerland
R. Manmatha, USA
Erik Mannens, Belgium
Stephane Marchand Maillet,
  Switzerland
Bernard Merialdo, France
Mark Maybury, USA
Vasileios Mezaris, Greece
Frank Nack, The Netherlands
Jan Nesvadba, New Zealand
Chung Wah Ngo, Hong Kong
Zeljko Obrenovic, The Netherlands
Noel O'Connor, Ireland
Stefan Rueger, UK
Lloyd Rutledge, The Netherlands
Ansgar Scherp, Ireland
Nicu Sebe, The Netherlands
Shin'ichi Satoh, Japan
Alan Smeaton, Ireland
Cees Snoek, The Netherlands
Hari Sundaram, USA
Vojtech Svatek, Czech Republic
Raphael Troncy, The Netherlands
Giovanni Tummarello, Ireland
Vassilis Tzouvaras, Greece
Mark van Doorn, The Netherlands
Jacco van Ossenbruggen,
  The Netherlands
Remco Veltkamp, The Netherlands
Paulo Villegas, Spain
Doug Williams, UK
Thomas Wittenberg, Germany
Marcel Worring, The Netherlands
Rong Yan, USA

Organizing Institution

Co-organizing Institutions

Conference held in Cooperation with

Sponsors

Deutsche
Forschungsgemeinschaft

Table of Contents

Keynote Talk: Tracking the Progress of Multimedia Semantics – from MPEG-7 to Web 3.0

Jane Hunter

The University of Queensland, Australia

Since 2001 when the original MPEG-7 standard was published, there have been a number of approaches to representing MPEG-7 as an ontology to enhance the semantic interoperability and richness of multimedia metadata. Subsequently range of tools/services, based on the MPEG-7 ontology, have been developed to generate both MPEG-7 and higher level semantic descriptions. In particular, a number of systems focused on semi-automatic semantic annotation of multimedia. These systems combine machine-learning techniques with manual ontology-based annotation to enhance the indexing and querying of multimedia content. However they depend on the availability of a training corpus of labelled multimedia content - which is difficult and expensive to generate. Most recently, there has been an explosion of social tagging tools on the Internet as part of the Web 2.0 phenomena. Such systems provide a community-driven approach to classifying resources on the Web, so that they can be browsed, discovered and re-used. Proponents of social tagging systems argue that they generate more relevant, light weight and cheaper metadata than traditional cataloguing systems. The future challenge is to leverage the potential of community-driven social tagging systems. This paper will describe the next generation of hybrid scalable classification systems that combine social tagging, machine-learning and traditional library classification approaches. It will also discuss approaches to the related challenge of aggregating light-weight community-generated tags with complex MPEG-7 descriptions and discipline-specific ontologies through common, extensible upper ontologies - to enhance discovery and re-use of multimedia content across disciplines and communities.

D. Duke et al. (Eds.): SAMT 2008, LNCS 5392, p. 1, 2008.

Keynote Talk: Data by the People, for the People

Mor Naaman

Rutgers University School of Communication, Information and Library Studies

What can we learn from social media and community-contributed collections of information on the web? The most salient attribute of social media is the creation of an environment that promotes user contributions in the form of authoring, curation, discussion and re-use of content. This activity generates large volumes of data, including some types of data that were not previously available. Even more importantly, design decisions in these applications can directly influence the users' motivations to participate, and hugely affect the resultant data. I will discuss the cycle of social media, and argue that a 'holistic' approach to social media systems, which includes design of applications and user research, can advance data mining and information retrieval systems.

Using Flickr as an example, I will describe a study in which we examine what motivates users to add tags and *geotags* to their photos. The new data enables extraction of meaningful (not to say *semantic*) information from the Flickr collection. We use the extracted information, for example, to produce summaries and visualizations of the Flickr collection, making the repository more accessible and easier to search, browse and understand as it scales. In the process, the user input helps alleviate previously intractable problems in multimedia content analysis.

D. Duke et al. (Eds.): SAMT 2008, LNCS 5392, p. 2, 2008.
© Springer-Verlag Berlin Heidelberg 2008

Automatic Summarization of Rushes Video Using Bipartite Graphs

Liang Bai[1,2], Songyang Lao[1], Alan F. Smeaton[2], and Noel E. O'Connor[2]

[1] Sch. of Information System & Management, National Univ. of Defense Technology,
ChangSha, 410073, R.P. China
lbai@computing.dcu.ie, laosongyang@vip.sina.com
[2] Centre for Digital Video Processing, Adaptive Information Cluster, Dublin City University,
Glasnevin, Dublin 9, Ireland
asmeaton@computing.dcu.ie, oconnorn@eeng.dcu.ie

Abstract. In this paper we present a new approach for automatic summarization of rushes video. Our approach is composed of three main steps. First, based on a temporal segmentation, we filter sub-shots with low information content not likely to be useful in a summary. Second, a method using maximal matching in a bipartite graph is adapted to measure similarity between the remaining shots and to minimize inter-shot redundancy by removing repetitive retake shots common in rushes content. Finally, the presence of faces and the motion intensity are characterised in each sub-shot. A measure of how representative the sub-shot is in the context of the overall video is then proposed. Video summaries composed of keyframe slideshows are then generated. In order to evaluate the effectiveness of this approach we re-run the evaluation carried out by the TREC, using the same dataset and evaluation metrics used in the TRECVID video summarization task in 2007 but with our own assessors. Results show that our approach leads to a significant improvement in terms of the fraction of the TRECVID summary ground truth included and is competitive with other approaches in TRECVID 2007.

Keywords: Video summarization, Evaluation.

1 Introduction

Decreasing capture and storage costs have led to significant growth in the amount and availability of video content in recent years. One consequence is that video summarization has recently emerged as an active research field. Video summaries provide a condensed version of a full-length video and should include the most important content from within the original video. Summaries can be used in different applications such as browsing and search, TV program editing, and so on. A variety of approaches have been proposed based on redundancy detection [1], frame clustering [2], speech transcripts [3], and multiple information streams [4]. Interest in this area has grown to such an extent that recently the TRECVID global benchmarking initiative initiated a work item on summarization, targeting rushes content i.e. extra video, B-rolls footage, etc. Rushes are captured by professional cameramen during the video production

D. Duke et al. (Eds.): SAMT 2008, LNCS 5392, pp. 3–14, 2008.

lifecycle. As an unedited version of the final video, they include many useless and redundant shots. Although the structure of the video and threading of the story are not directly available, the rushes are organized based on the traditional shot structure.

In 2007, the National Institute of Standards and Technology (NIST) coordinated an evaluation of automatic video summarization for rushes. This took place as part of the larger video benchmarking activity known as TRECVID. The overall video summarization task, data used, evaluation metrics, etc., are described in [5]. Importantly, in the TRECVID guidelines for rushes summarization, several criteria are used for evaluating the generated summaries, including the fraction of objects and events included by the summary (IN), the ease of understanding the summary (EA), the time needed for subjective judgment (TT, VT), and the compactness of the summary (DU, XD).

For our participation in this task, we proposed a relatively straightforward key-frame-based approach [1]. However, this approach did not perform as well as expected, especially in the IN and EA criteria. The inclusion results placed our approach (mean: 0.38; median: 0.38) among the 5 lowest scoring participants. Our low EA scores (mean: 2.53; median: 2.67) placed us second worst out of 25 participants. This poor performance encouraged us to undertake detailed failure analysis and motivated us to re-analyze the characteristics of rushes videos. This paper reports on the new algorithm that we developed adapted from [1] on the basis of the results of this analysis.

There are two types of redundant information in rushes video. The first is content such as clapperboards, color bars, monochromatic shots and very short shots. This content is not related to the main content of the video and is not of value in a summary. The second type of redundant content is repetition of some shots with near-identical material appearing in the second and subsequent shots. During program production, the same shot is often taken many times. For summarization purposes, retake shots should be detected and only one kept, removing others from the final summary.

Our enhanced approach described in this paper focuses on representative frames selection, useless content removal, retake detection and content filtering and ranking in the selected shots. In order to select representative frames, which represent video content with as much precision as possible, we calculate the difference between consecutive frames based on color features at the pixel level in each shot and use a geometrical method to select representative frames. Although we don't explicitly segment sub-shots, our method for key frame selection guarantees that representative frames in each sub-shot are selected as both the sum of differences and length of the shot are considered. SVM classifiers are trained based on the TRECVID development data to detect color bars and monochromatic frames. Clapperboard clips are removed by an existing method for Near-Duplicate Keyframe (NDK) detection. After filtering the useless content, we reduce the inter-shot redundancy by removing repeated retake-shots. Maximal matching based on the Hungarian algorithm is then adopted to measure the similarity between retake-shots at the level of key-frames. Finally, we reduce the intra-shot redundancy of the remaining shots in two steps:

1. We remove similar sub-shots by calculating the color similarity between key-frames that represent sub-shots;
2. We detect the important content including the presence of a face and motion intensity to score remaining key-frames and keep the key-frames with higher score according to the time limitation requirements of the final summary.

The key difference between the approach presented in this paper and our original described in [1], is the introduction of maximal matching in a bipartite graph to measure similarity between shots and this is the reason for the significantly improved performance reported in this paper. Figure 1 describes our overall approach to rushes summarization. First, a given rushes video is structured into shots and sub-shots and useless sub-shots are filtered (see Section 2 and Section 3). Then, inter-shot redundancy is reduced by removing repetitive re-take shots (see Section 4). Finally, a measure is proposed to score the presence of faces and motion for intra-shot redundancy removal (see Section 5). We present a summary of our experimental results in Section 6 and some conclusions in Section 7.

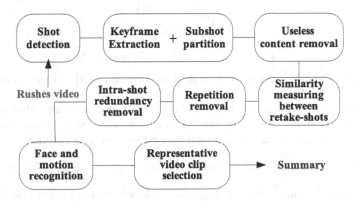

Fig. 1. Our approach to rushes video summarization

2 Video Structuring

Given the raw nature of rushes video, we first structure it by detecting shots and sub-shots and extracting key-frames from each sub-shot. Since all rushes videos are unedited, hard cuts typically dominate the transitions used and so we focus only on detection of hard cuts. In our work we use a mutual information measure between two successive frames calculated separately for each RGB channel. The mutual information between two successive frames is calculated separately for each of the R, G and B channels. In the case of the R component, the element $C_{l,l+1}^{R}(i, j), 0 \le i, j \le N-1$, N being the number of gray levels in the image, corresponds to the probability that a pixel with gray level i in frame f_t has gray level j in frame f_{t+1}. The mutual information of frame f_k, f_l for the R component is expressed as:

$$I_{k,l}^{R} = -\sum_{i=0}^{N-1}\sum_{j=0}^{N-1} C_{k,l}^{R}(i, j) \log \frac{C_{k,l}^{R}(i, j)}{C_{k}^{R}(i)C_{l}^{R}(j)}$$

The total mutual information between frames f_k and f_l is defined as:

$$I_{k,l} = I_{k,l}^{R} + I_{k,l}^{G} + I_{k,l}^{B}$$

A smaller value of the mutual information leads to a high probability of a large difference in the content between two frames. Local mutual information mean values on a temporal window W of size N_w for frame f_t are calculated as:

$$\bar{I}_t = \frac{\sum_{i=t}^{N_w+t} I_{i,i+1}}{N_w}$$

The standard deviation of mutual information on the window is calculated as:

$$\sigma_I = \sqrt{\frac{\sum_{i=t}^{N_w+t} \left(I_{i,i+1} - \bar{I}_t\right)^2}{N}}$$

The quantity $\dfrac{\left|\bar{I}_t - I_{t,t+1}\right|}{\sigma_I}$ is then compared to a threshold H, which represents the mutual information variation at frame f_t deviating from the mean value and determines a boundary frame. Assuming that the video sequence has a length of N frames, the shot boundary determination algorithm may be summarized as follows:

Step 1: calculate the mutual information time series $I_{t,t+1}$ with $0 \leq t \leq N - N_w$.

Step 2: calculate $\bar{I}_t$ and σ_I at each temporal window in which f_t is the first frame.

Step 3: if $\dfrac{\left|\bar{I}_t - I_{t,t+1}\right|}{\sigma_I} \geq H$, frame f_t is determined as a shot boundary.

We evaluated the effectiveness of this on the TRECVID development data that is provided for training and familiarization purposes and achieved an overall performance of 93.4% recall and 91.5% precision, which is acceptably close to the state of the art.

In rushes video, each shot usually contains not only the scripted action, but also other material that is not related to the story, such as camera adjustments, discussions between the director and actors, and unintentional camera motion. Further, the scripted action usually contains varied content because of camera and/or object movements. In video summarization, we aim to remove video segments not related to the story and to include only the other video segments. One key-frame for each shot, however, is not enough for this purpose and so we partition each shot into sub-shots corresponding to different content.

We split each frame into an 8x8 pixel grid and calculate the mean and variance of RGB color in each grid. The Euclidean distance is then used to measure the difference between neighboring frames. Usually, in one sub-shot the cumulative frame difference shows gradual change. High curvature points within the curve of the cumulative frame difference are very likely to indicate the sub-shot boundaries. Figure 2 explains this idea. After sub-shot partitioning, the key-frames are selected at the midpoints between two consecutive high curvature points.

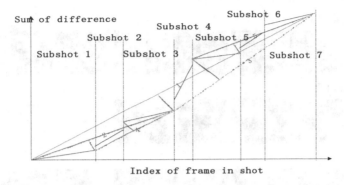

Fig. 2. Sub-shots partition

3 Useless Content Removal

Samples of useless content contained in rushes video are illustrated in Figure 3. These include color bars, monochromatic shots, clapperboards and short shots. First, shots of duration less than 1 second are removed. For color bars and monochromatic shots, four features including color layout, scalable color, edge histogram and homogenous texture are extracted from the key-frames in the corresponding shots. SVM classifiers are trained to recognize color bars and monochromatic shots.

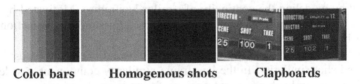

Color bars   Homogenous shots   Clapboards

Fig. 3. Examples of useless content

We employ the algorithm for Near-Duplicate Keyframe (NDK) detection described in [6] to detect clapperboards. A set of 50 example keyframes of the clapboards were extracted from the TRECVID development set. The regions where clapperboards are present are manually annotated. Among the keyframes of each shot in the given rushes video, we detect the keypoints and match them with the example clapperboards. If enough matches are found that lie in the annotated regions, the keyframe is detected as a clapperboard and removed.

4 Re-take Shot Detection and Removal

In rushes video, the same shot can be re-taken many times in order to eliminate actor or filming mistakes. In this case, the re-take shots should be detected and the most satisfactory ones kept, removing the others from the final summarization. Rows 1, 2 and 3 in Figure 4 show the keyframes extracted from three retake-shots in one of the rushes test videos.

Retake-shot 1

Retake-shot 2

Retake-shot 3

A different shot

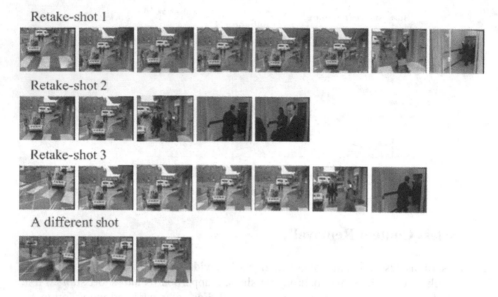

Fig. 4. Examples of retake-shots

We assume that the similarity between shots can be measured according to the similarity of keyframes extracted from corresponding shots. Thus, the re-take shots are detected by modeling the continuity of similar keyframes. Motivated by maximal matching in bipartite graphs, an approach is proposed for similarity detection between video shots based on this matching theory.

Our similarity measure between video shots is divided into two phases: key frame similarity and shot similarity. In the key frame similarity component, a video shot is partitioned into several sub-shots and one key frame is extracted from each sub-shot. The similarity among sub-shots is used instead of the similarity between corresponding key frames. Key frame similarity is measured according to the spatial color histogram and texture features.

A shot can be expressed as: $S = \{k_1, k_2, ..., k_n\}$, where k_i represents the ith keyframe. So, for two shots, $Sx = \{kx_1, kx_2, ..., kx_n\}$ and $Sy = \{ky_1, ky_2, ..., ky_m\}$, the similar keyframes between Sx and Sy can be expressed by a bipartite graph $G = \{Sx, Sy, E\}$,

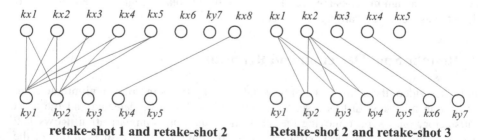

retake-shot 1 and retake-shot 2   **Retake-shot 2 and retake-shot 3**

Fig. 5. Two examples of bipartite graph for retake-shots

where $V = Sx \cup Sy$, $E = \{e_{ij}\}$, e_{ij} indicates kx_i is similar to ky_j. Figure 5 illustrates two examples of bipartite graphs for retake-shot 1, retake-shot 2 and retake-shot 3 shown in Figure 4.

Clearly, there exist many similar pairs of keyframes between two retake-shots. But in our experiments we also find there often exist similar keyframes in one retake-shot. This results in one to many, many to one and many to many relations in a bipartite graph. In this case, there will be many similar keyframes pairs found between two dissimilar shots. The bipartite graph between retake-shot 3 and a different shot shown in Figure 4 illustrates this case in Figure 6.

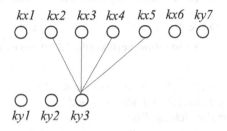

Fig. 6. A bipartite graph between two dissimilar shots

If we use the number of similar keyframe pairs to determine the retake-shots, 4 similar keyframe pairs are found in the Sx shot shown in Figure 6 and this exceeds half of the keyframes in Sx. In this case, Sx is likely to be determined as similar to Sy, whilst this is not the case in practice.

In our approach, the similarity between two shots is measured by the maximal matching of similar keyframes in the bipartite graph model. The Hungarian algorithm

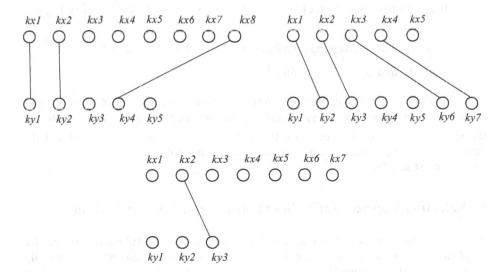

Fig. 7. Examples of maximal matching results

[7] is used to calculate maxima matching M, $M \subseteq E$. If $M \geq \min\{\left\lceil\frac{2}{3}n\right\rceil, \left\lceil\frac{2}{3}m\right\rceil\}$

where n,m are the number of keyframes in these two shots. These thresholds were chosen based on experimental results in order to give the best similarity matches. Figure 7 shows the maximal matching results of the examples shown in Figure 5 and Figure 6.

From Figure 7, we can find the maximal matching of dissimilar shots is 1. From this, it should be clear that it is relatively straightforward to determine the true retake-shots according to the maximal matching.

The matching steps using the Hungarian algorithm are as follows:

Assumption: A given bipartite graph is $G_k = \{Sx, Sy_k, E_k\}$; "0" denotes a vertex that is not searched, "1" denotes a saturation vertex and "2" denotes a vertex that cannot increase the matching.

Step1: Given an initial matching M, mark the vertexes "1";
Step2: Check if every vertex in Sx has a non-"0" mark.
    If yes, M is the maximal matching. End.
    If no, find a vertex marked "0" $x_0 \in Sx$, let $A \leftarrow \{x_0\}, B \leftarrow \phi$.
Step3: Check if $N(A) = B(N(A) \subseteq Sy_k)$. $N(A)$ denotes the vertices belonging to Sy_k that neighbor with the vertices in A. $B(N(A) \subseteq Sy_k)$ denotes the vertices belonging to Sx_k that neighbor the vertices in $N(A)$.

    If yes, x_0 cannot increase matching, mark x_0 "2", go to Step2;

    If no, find a vertex Sy_i in $N(A) - B$, check if Sy_i is marked with "1".

        If yes, there exists an edge $(Sy_i, z) \in M$, let $A \leftarrow A \cup \{z\}, B \leftarrow B \cup \{Sy_i\}$, go to Step3.

        If no, exist an augmenting path from x_0 to Sy_i, let $M \leftarrow M \oplus P$,

    mark x_0 and Sy_i "1", go to Step2.

The complexity of this algorithm is $O(ne)$, where n is the number of vertices of Sx in the bipartite graph $G = \{Sx, Sy, E\}$ and e is the number of edges. After measuring the similarity of shots, re-take shots are detected. The last shot is retained and the others are removed since in rushes content the last retake shot is usually the one deemed most satisfactory.

5 Selecting Representative Shots and Summary Generation

After low value content and repetitive re-take shot removal, useful content is kept for summary generation. However the volume of the remaining content still typically exceeds the useful duration limit set by the TRECVID guideline — set at 4% duration of the original video in 2007. So, the most representative clips need to be selected to generate the final summary. In our work, we extract motion and face factors to rank how representative each remaining sub-shot is in the context of the overall video.

A three-stage process, achieved using the aceToolbox [8], is used to describe the level of motion activity in each sub-shot. First, MPEG-1 motion vector data is extracted from the video. Next, the percentage of non-zero blocks in the frame (where a high percentage indicates higher motion activity) is calculated for each frame in the video. Finally, this per-frame data is used along with the shot-boundary data calculated previously to compute an average motion measure for the entire sub-shot. As a result, each keyframe in a given sub-shot is assigned the same measure of motion activity.

Our face detection processing extends the Bayesian Discriminating Feature (BDF) originally proposed by Liu [9] for detecting frontal faces in grayscale images. Using a statistical skin color model [10], we can detect multiple faces at various sizes and orientations within color images. Ideally this processing would be carried out for each frame of the original footage, however, for efficiency we only perform this operation on the detected keyframes. While this potentially results in the loss of information, such as the prevalence of faces across shots, it ensures efficient processing while still providing enough information to reliably enhance summary construction.

Sub-shot duration is important for sub-shot selection so we use simple weighting to combine the factors.

$$Score = (Number-of-faces\,/\,Maximum-faces-in-Footage\times0.3)$$
$$+(Amount-of-motion\times0.3)$$
$$+(Duration-of-subshot\,/\,Total-Duration-of-All\times0.4)$$

These weightings for the different components were chosen as discussed previously in [1]. Once the representative scores for sub-shots are calculated, the sub-shots with highest scores are selected according to the summary duration limitation. Finally, 1-second clips centred around the keyframe in each selected sub-shot are extracted for generating our final summary.

6 Experiments Results

Using our approach, we generated the summaries for all test rushes videos. The seven criteria set by the TRECVID guidelines for summarization evaluation are:

- EA: Easy to understand: (1 strongly disagree - 5 strongly agree);
- RE: Little duplicate video: (1 strongly disagree - 5 strong agree);
- IN: Fraction of inclusions found in the summary (0 - 1);
- DU: Duration of the summary (sec);
- XD: Difference between target and actual summary size (sec);
- TT: Total time spent judging the inclusions (sec);
- VT: Video play time (vs. pause) to judge the inclusions (sec).

IN, DU and XD are objective criteria that we can calculate directly from the TRECVID groundtruth to evaluate our summaries. However, EA, RE, TT and VT are criteria that depend on subjective judgments. Thus for a complete evaluation of our proposed approach it was necessary to re-run the evaluation performed by NIST with our own test subjects. Ten participants (all students in the School of Information System & Management, National University of Defense Technology) were selected to review the summaries under the exact same guidelines as provided by NIST and give their score for the four subjective criteria.

Of course, by running our own evaluation we could potentially introduce new subjective variations into the evaluation process. To investigate this, we first evaluated three sets of results: the two TRECVID baselines (see [5] for details) and our own original submission. The experimental results we obtained are compared to the official results reported from TRECVID in Table 1.

The results in Table 1 show that there exists a small difference in the subjective judgments between our participants and NIST assessors. This is understandable given that different people have different skills, intellects, powers of discernment, etc. However, from Table 1 we can see that the difference of judgments between our participants and NIST assessors is small. From this we conclude that our participants' evaluations on the subjective criteria are reasonable and credible. Given this, we proceeded to re-run the complete evaluation.

Table 1. Experimental results for the comparison between our participants and NIST assessors

Criterion		EA	RE	TT	VT
TrecBaseline1	Our Participants	3.12	3.26	115.45	73.20
	NIST	3.33	3.33	110.67	66.67
TrecBaseline2	Our Participants	3.35	3.30	118.10	70.38
	NIST	3.67	3.67	109.17	63.83
Our original [2]	Our Participants	2.29	3.33	76.78	48.49
	NIST	2.67	3.67	70.83	42.67

Table 2. Experiment results for IN, DU and XU

Criterion	IN	DU	XD
TRECVID Baseline1	0.60	66.40	-2.28
TRECVID Baseline2	0.62	64.60	-0.89
Mean of all 22 teams	0.48	49.54	10.33
Our original [2]	0.38	40.90	8.65
Our enhanced	**0.78**	**41.61**	**18.83**

Table 3. Experiment results for EA, RE, TT and VT

Criterion	EA	RE	TT	VT
TRECVID Baseline1	3.12	3.26	115.45	73.20
TRECVID Baseline2	3.35	3.30	118.10	70.38
Our original [2]	2.29	3.33	76.78	48.49
Our enhanced	**3.74**	**3.88**	**89.21**	**44.50**

The experimental results for all of our summaries are shown in Table 2 and Table 3. The results in Table 2 show that our enhanced approach results in a big improvement in IN (0.40) with a slightly longer duration of summaries (0.71 sec) compared with our original approach. Of particular note is the fact that our enhanced approach's XD is 18.83, which is 8.5 sec longer than the mean of the other 22 teams. This is because we tend to retain the valuable content in rushes as much as possible within the summary duration constraint. Table 3 shows the evaluation results for the four subjective criteria. Clearly we obtain very encouraging results for the EA and RE. These experimental results clearly show that our enhanced approach performs competitively compared with the other teams and the baselines.

7 Conclusion and Discussion

This paper describes our approach to summarizing rushes video content. It focuses on the adaptation of an approach we used in the TRECVID summarization task. In this approach, we employ shot and sub-shot detections for video structuring and we train SVMs for removing useless content. We model the similarity of keyframes between two shots by bipartite graphs and we measure shot similarity by maximal matching for re-take shot detection. Based on consideration of motion, face and duration, sub-shots are ranked and the most representative clips are selected for inclusion in the final summary. This key different with respect to our original TRECVID submission is the inclusion of bipartite matching. To evaluate this new approach, we re-ran the evaluation procedure ourselves with our own assessors. Experimental results indicate that the subjective evaluation is in line with that originally carried out by NIST. Our improved approach clearly demonstrates improvements compared to our original approach, but more importantly compared to the TRECVID baselines and the other teams who participated.

Not withstanding this, the summarization problem clearly still remains challenging. Indeed, most submissions cannot significantly outperform the two baselines, which are simply based on fixed-length shot selection and visual clustering. This poses the key question as to whether a deeper semantic understanding of the content can help in this regard.

Acknowledgments. This work is supported by the National High Technology Development 863 Program of China (2006AA01Z316), the National Natural Science Foundation of China (60572137) and Science Foundation Ireland through grant 03/IN.3/I361.

References

1. Byrne, D., Kehoe, P., Lee, H., O Conaire, C., Smeaton, A.F., O'Connor, N., Jones, G.: A User-Centered Approach to Rushes Summarisation Via Highlight-Detected Keyframes. In: Proceedings of the TRECVID Workshop on Video Summarization (TVS 2007), Augsburg, Germany, September 28, pp. 35–39. ACM Press, New York (2007)
2. Ferman, A.M., Tekalp, A.M.: Two-stage hierarchical video summary extraction to match low-level user browsing preferences. IEEE Trans. Multimedia 5(2), 244–256 (2003)

3. Ma, Y.F., Lu, L., Zhang, H.J., Li, M.: A User Attention Model for Video Summarization. In: Proc. ACM Multimedia Conference (2002)
4. Taskiran, C.M., Pizlo, Z., Amir, A., Ponceleon, D., Delp, E.: Automated Video Program Summarization Using Speech Transcripts. IEEE Trans. on Multimedia 8(4), 775–791 (2006)
5. Over, P., Smeaton, A.F., Kelly, P.: The TRECVID 2007 BBC rushes summarization evaluation pilot. In: Proceedings of the TRECVID Workshop on Video Summarization (TVS 2007), Augsburg, Germany, September 28, 2007, pp. 1–15. ACM Press, New York (2007)
6. Ngo, C.W., Zhao, W.L., Jiang, Y.G.: Fast Tracking of Near-Duplicate Keyframes in Broadcast Domain with Transitivity Propagation. In: Proc. ACM Multimedia Conference (October 2006)
7. Dai, Y.Q., Hu, G.Z., Chen, W.: Graph Theory and Algebra Structure, pp. 89–91. Tsinghua University Press, Beijing (1995) (in Chinese)
8. O'Connor, N., Cooke, E., le Borgne, H., Blighe, M., Adamek, T.: The AceToolbox: Low-Level Audiovisual Feature Extraction for Retrieval and Classification. In: Proceedings 2nd IEE European Workshop on the Integration of Knowledge, Semantic and Digital Media Technologies, London, U.K (2005)
9. Liu, C.: A Bayesian discriminating features method for face detection. IEEE Tran. on PAMI 25, 725–740 (2003)
10. Cooray, S., O'Connor, N.: A Hybrid Technique for Face Detection in Color Images. In: IEEE Conf. on Advanced Video Surveillance (AVSS 2005), Italy, September 15-16 (2005)

Validating the Detection of Everyday Concepts in Visual Lifelogs

Daragh Byrne[1,2], Aiden R. Doherty[1,2], Cees G.M. Snoek[3], Gareth G.F. Jones[1], and Alan F. Smeaton[1,2]

[1] Centre for Digital Video Processing, Dublin City University, Glasnevin, Dublin 9, Ireland
[2] CLARITY: Centre for Sensor Web Technologies
{daragh.byrne,aiden.doherty,gareth.jones,
alan.smeaton}@computing.dcu.ie
[3] ISLA, University of Amsterdam, Kruislaan 403, 1098SJ Amsterdam, The Netherlands
cgmsnoek@ uva.nl

Abstract. The Microsoft SenseCam is a small lightweight wearable camera used to passively capture photos and other sensor readings from a user's day-to-day activities. It can capture up to 3,000 images per day, equating to almost 1 million images per year. It is used to aid memory by creating a personal multimedia lifelog, or visual recording of the wearer's life. However the sheer volume of image data captured within a visual lifelog creates a number of challenges, particularly for locating relevant content. Within this work, we explore the applicability of semantic concept detection, a method often used within video retrieval, on the novel domain of visual lifelogs. A concept detector models the correspondence between low-level visual features and high-level semantic concepts (such as indoors, outdoors, people, buildings, etc.) using supervised machine learning. By doing so it determines the probability of a concept's presence. We apply detection of 27 everyday semantic concepts on a lifelog collection composed of 257,518 SenseCam images from 5 users. The results were then evaluated on a subset of 95,907 images, to determine the precision for detection of each semantic concept and to draw some interesting inferences on the lifestyles of those 5 users. We additionally present future applications of concept detection within the domain of lifelogging.

Keywords: Microsoft SenseCam, lifelog, passive photos, concept detection, supervised learning.

1 Introduction

Recording of personal life experiences through digital technology is a phenomenon we are increasingly familiar with: music players, such as iTunes, remember the music we listen to frequently; our web activity is recorded in web browsers' "History"; and we capture important moments in our life-time through photos and video [1]. This concept of digitally capturing our memories is known as lifelogging. Lifelogging and memory capture was originally envisaged to fulfill at least part of Vannevar Bush's 1945 MEMEX vision. Bush describes his MEMEX as a collection in which a person

D. Duke et al. (Eds.): SAMT 2008, LNCS 5392, pp. 15–30, 2008.
© Springer-Verlag Berlin Heidelberg 2008

could store all of their life experience information including photographs, documents and communications *"and which is mechanized so that it may be consulted with exceeding speed and flexibility."* [3].

A visual lifelogging device, such as the SenseCam, will capture approximately 1 million images per year. The sheer volume of photos collected, and the rate at which a collection can grow, pose significant challenges for the access, management, and utility of such a lifelog. However, inroads to resolving some of the concerns relating to these issues have already been made. For example, in prior work we proposed the aggregation of individual images within a visual lifelog into higher level discrete 'events' which represent single activities in a user's day [8]. Furthermore, work has been carried out to investigate how best to select a single representative keyframe image which best summarises a given event [7]. Lee *et. al.* have constructed an event-oriented browser which enables a user to browse each day in their collection through a calendar controlled interface [19]. This interface allows the 'gisting' or recap of an entire day's activities by presenting a visual summary of the day. The benefit of such daily summaries has been highlighted in the results of a preliminary study carried out between Microsoft Research and Addenbrooke's Hospital, Cambridge, U.K. where visual lifelog recordings notably improved subjects' recall of memories [15].

A fundamental requirement outlined in Bush's MEMEX [3] is that we must provide on-demand, rapid and easy access to the memories and experiences of interest and to achieve this we must be able to support high quality retrieval. While many steps have been taken towards managing such an ever-growing collection [7,8,20], we are still far from achieving Bush's original vision. This is mainly due to the fact that we cannot yet provide rapid, flexible access to content of interest from the collection.

The most obvious form of content retrieval is to offer refinement of the lifelog collection based on temporal information. Retrieval may also be enabled based on the low-level visual features of a query image. However, in order for such a search to be effective the user must provide a visual example of the content they seek to retrieve and there may be times when a user will not possess such an example, or that it may be buried deep within the collection. Augmentation and annotation of the collection with sources of context metadata is another method by which visual lifelogs may be made searchable. Using sources of context such as location or weather conditions has been demonstrated to be effective in this regard [4,10]. There are, however, limitations to these approaches as well, most importantly any portion of the collection without associated context metadata would not be searchable. Moreover, while information derived from sensors such as Bluetooth and GPS [4] may cover the 'who' and the 'where' of events in an individual's lifelog, however, they do not allow for the retrieval of relevant content based on the 'what' of an event.

An understanding of the 'what' or the semantics of an event would be invaluable within the search process and would empower a user to rapidly locate relevant content. Typically, such searching is enabled in image tools like Flickr through manual user contributed annotations or 'tags', which are then used to retrieve visual content. Despite being effective for retrieval, such a manual process could not be practical within the domain of lifelogging, since it would be far too time and resource intensive given the volume of the collection and the rate at which it grows. Therefore we should explore methods for automatic annotation of visual lifelog collections.

One such method is concept detection, an often employed approach in video retrieval [22,24,27], which aims to describe visual content with confidence values indicating the presence or absence of object and scene categories. Although it is hard to bridge the gap between low-level features that one can extract from visual data and the high-level conceptual interpretation a user gives to this data, the video retrieval field has made substantial progress by moving from specific single concept detection methods to generic approaches. Such generic concept detection approaches are achieved by fusion of color-, texture-, and shape-invariant features [11,12,14,25], combined with supervised machine learning using support vector machines [5,26]. The emphasis on generic indexing by learning has opened up the possibility of moving to larger concept detector sets [16,23,28]. Unfortunately these concept detector sets are optimized for the (broadcast) video domain only, and their applicability to other domains such as visual lifelog collections remains as of yet unclear.

Visual lifelog data, and in particular Microsoft SenseCam data – the source for our investigation - is markedly different from typical video or photographic data and as such presents a significantly more challenging domain for visual analysis. SenseCam images tend to be of low quality owing to: their lower visual resolution; their use of a fisheye lens which distorts the image somewhat but increases the field of view; and a lack of flash resulting in many images being much darker than desired for optimal visual analysis. Also, almost half of the images are generally found to contain non-desirable artifacts such as grain, noise, blurring or light saturation [13]. Thus our investigation into the precision and reliability of semantic concept detection methods will provide important insights into their application for visual lifelogs.

The rest of this paper is organised as follows: Section 2 details how we apply concept detection to images captured by the SenseCam lifelogging device; section 3 quantitatively describes how accurate our model is in detecting concepts; section 4 provides interesting inferences on the lifestyles of our users using the detected concepts; while sections 5 and 6 finally summarise this work and detail many interesting future endeavours to be investigated.

2   Concept Detection Requirements in the Visual Lifelog Domain

The major requirements for semantic concept detection on visual lifelogs are as follows: 1) Identification of Everyday Concepts; 2) Reliable and Accurate Detection; and 3) Identification of Positive and Negative Examples. We now discuss how we followed these steps with respect to lifelog images captured by a SenseCam.

Use Case: Concept Detection in SenseCam Images
To study the applicability of concept detection in the lifelog domain we make use of a device known as the SenseCam. Microsoft Research in Cambridge, UK, have developed the SenseCam as a small wearable device that passively captures a person's day-to-day activities as a series of photographs and readings from in-built sensors [15]. It is typically hung from a lanyard around the neck and, so is oriented towards the majority of activities which the user is engaged in. Anything in the view of the wearer can be captured by the SenseCam because of its fisheye lens. At a minimum the SenseCam will automatically take a new image approximately every 50 seconds, but

sudden changes in the environment of the wearer, detected by onboard sensors, can trigger more frequent photo capture. The SenseCam can take an average of 3,000 images in a typical day and, as a result, a wearer can very quickly build large and rich photo collections. Already within a year, the lifelog photoset will grow to approximately 1 million images!

Fig. 1. The Microsoft SenseCam (Inset: right as worn by a user)

2.1 Collection Overview

In order to appropriately evaluate concept detection we organised the collection of a large and diverse dataset of 257,518 SenseCam images gathered by five individual users. In order to further ensure diversity, there was no overlap between the periods captured within each user's dataset. A breakdown of the collection is illustrated in Table 1. It is worth noting that not all collections featured the same surroundings. Often collections were subject to large changes in location, behaviour, and environments. This allowed us to more reliably determine the robustness of concept detection in this domain.

Table 1. An overview of the image collection used

User	Total Images	Number of Concepts Annotated	Days Covered
1	79,595	2,180	35
2	76,023	9,436	48
3	42,700	27,223	21
4	40,715	28,023	25
5	18,485	11,408	8

2.2 Determining LifeLog Concepts

Current approaches to semantic concept detection are based on a set of positive and a set of negative labeled image examples by which a classifier system can be trained (see section 2.3). Consequently, as part of this investigation we identify the concepts within the collection for which a set of training examples would be collected. In order to determine the everyday concepts within the collection, a subset of each user's collection was visually inspected by playing the images sequentially at highly accelerated speed. A list of concepts previously used in video retrieval [22,23] and agreed upon as applicable to a SenseCam collection were used as a starting point. As a new identifiable 'concept' was uncovered within the collection it was added to this list. Each observed repetition of the concept gave it additional weight and ranked it more

Table 2. An outline of the 27 concepts and the no. of positive examples per concept and per user

Concept / User	1	2	3	4	5	All
Total to Annotate	16111	14787	8593	8208	3697	51396
Indoors	1093	1439	6790	6485	3480	19287
Hands	1	17	4727	3502	2402	10649
Screen (computer/laptop)	7	1101	4699	2628	2166	10601
Office	7	78	4759	2603	336	7783
People	0	1775	573	3396	889	6633
Outdoors	250	915	1248	812	67	3292
Faces	0	553	101	1702	662	3018
Meeting	0	808	0	1233	355	2396
Inside of vehicle, not driving (airplane, car, bus)	257	1326	420	223	0	2226
Food (eating)	0	795	349	870	129	2143
Buildings	140	49	981	621	62	1853
Sky	0	202	720	525	66	1513
Road	125	0	231	648	4	1008
Tree	24	44	378	469	42	957
Newspaper/Book (reading)	0	85	13	520	309	927
Vegetation	0	3	255	468	52	778
Door	28	0	279	128	144	579
Vehicles (external view)	33	0	322	121	4	480
Grass	0	122	99	190	33	444
Holding a cup/glass	0	0	21	353	44	418
Giving Presentation / Teaching	0	43	0	309	0	352
Holding a mobile phone	0	4	54	28	147	233
Shopping	0	75	102	48	3	228
Steering wheel (driving)	208	0	0	0	0	208
Toilet/Bathroom	6	0	75	93	0	174
Staircase	0	2	26	48	11	87
View of Horizon	1	0	1	0	1	3

highly for inclusion. Over 150 common concepts were identified in this process. It was decided that the most representative (i.e. everyday) concepts should be selected and as such these were then narrowed to just 27 core concepts through iterative review and refinement. Criteria for this refinement included the generalisability of the concept across collections and users. For example, the concepts 'mountain' and 'snow' occurred in User 1's collection frequently but could not be considered as an everyday concept as it was not present in the remaining collections. As such the 27

Fig. 2. Visual examples of each of the 27 everyday concepts as detected and validated for the lifelog domain

concepts represent a set of everyday core concepts most likely to be collection independent, which should consequently be robust with respect to the user and setting. These core concepts are outlined in Figure 2 using visual examples from the collection. Given that some concepts are related (e.g. it is logical to expect that 'buildings' and 'outdoors' would co-occur), it is important to note that each image may contain multiple (often semantically related) concepts.

A large-scale manual annotation activity was undertaken to provide the required positive and negative labeled image examples. As annotating the entire collection was impractical and given that SenseCam images tend to be temporally consistent the collection was skimmed by taking every 5th image. Collection owners annotated their own SenseCam images for the presence of each of the concepts. As by their nature lifelog images are highly personal, it is important for privacy reasons that it is only the owner of the lifelog images who labels his or her images. Therefore, collection owners annotated their own SenseCam images for each concept. This also provided the opportunity for them to remove any portion of their collection they did not wish to have included as part of this study. All users covered their entire skimmed collection with the exception of User 1, who only partially completed the annotation process on a subset of his collection. The number of positive examples for each concept and for each user is presented in Table 2.

2.3 Concept Detection Process

Our everyday concept detection process is composed of three stages: 1) supervised learning, 2) visual feature extraction, and 3) feature and classifier fusion, each of these stage uses the implementation detailed below.

Supervised Learner: We perceive concept detection in lifelogs as a pattern recognition problem. Given pattern x, part of an image i, the aim is to obtain a probability measure, which indicates whether semantic concept ω_j is present in image i. Similar to [16,24,27,28], we use the Support Vector Machine (SVM) framework [26] for supervised learning of concepts. Here we use the LIBSVM implementation [5] with radial basis function and probabilistic output [21]. We obtain good SVM settings by using an iterative search on a large number of parameter combinations.

Visual Feature Extraction: For visual feature extraction we adopt the well-known codebook model, see e.g. [17], which represents an image as a distribution over codewords. We follow [25] to build this distribution by dividing an image in several overlapping rectangular regions. We employ two visual feature extraction methods to obtain two separate codebook models, namely: 1) *Wiccest features,* which rely on natural image statistics and are therefore well suited to detect natural sceneries, and 2) *Gabor features,* which are sensitive to regular textures and color planes, and therefore well suited for the detection of man-made structures. Both these image features measure colored texture.

Wiccest features [11] utilise natural image statistics to model texture information. Texture is described by the distribution of edges in a certain image region. Hence, a histogram of a Gaussian derivative filter is used to represent the edge statistics. It was shown in [12] that the complete range of image statistics in natural textures can be well modeled with an integrated Weibull distribution, which in turn can be characterised by just 2 parameters. Thus, 2 Weibull parameter values for the x-edges and y-edges of the three color channels yields a 12 dimensional descriptor. We construct a codebook model from this low-level region description by computing the similarity between each region and a set of 15 predefined semantic color-texture patches (including e.g. sand, brick, and water), using the accumulated fraction between their Weibull parameters as a similarity measure [25]. We perform this procedure for two region segmentations, two scales, the x- and the y-derivatives, yielding a codebook feature vector of 120 elements we term w.

Gabor filters may be used to measure perceptual surface texture in an image [2]. Specifically, Gabor filters respond to regular patterns in a given orientation on a given scale and frequency. In order to obtain an image region descriptor with Gabor filters we follow these three steps: 1) parameterise the Gabor filters, 2) incorporate color invariance, and 3) construct a histogram. First, the parameters of a Gabor filter consist of orientation, scale and frequency. We use four orientations, 0°, 45°, 90°, 135°, and two (scale, frequency) pairs: (2.828, 0.720), (1.414, 2.094). Second, color responses are measured by filtering each color channel with a Gabor filter. The W color invariant is obtained by normalising each Gabor filtered color channel by the intensity [14]. Finally, a histogram is constructed for each Gabor filtered color channel. We construct a codebook model from this low-level region description by again computing the similarity between each region and a set of 15 predefined semantic color-texture patches, where we use histogram intersection as the similarity measure. Similar to the procedure for w, this yields a codebook feature vector of 120 elements we term g.

Feature and Classifier Fusion: As the visual features w and g emphasise different visual properties, we consider them independent. Hence, much is to be expected from their fusion. We employ fusion both at the feature level as well as the classifier level. Although the vectors w and g rely on different low-level feature spaces, their codebook model is defined in the same codeword space. Hence, for feature fusion we can concatenate the vectors w and g without the need to use normalisation or transformation methods. This concatenation yields feature vector f.

For each of the feature vectors in the set $\{w, g, f\}$ we learn a supervised classifier. Thus for a given image i and a concept ω_j, we obtain three probabilities, namely: $p(\omega_j | w_i)$, $p(\omega_j | g_i)$, and $p(\omega_j | f_i)$, based on the same set of labeled examples. To maximize the impact of our labeled examples, we do not rely on supervised learning in the classifier fusion stage. Instead, we employ average fusion of classifier probability scores, as used in many visual concept detection methods [16,24,27,28]. After classifier fusion we obtain our final concept detection score, which we denote $p(\omega_{ij})$.

3   Validation of Everyday Concept Detection

In order to validate $p(\omega_{ij})$, we manually judged a subset of the collection. To make a determination of its presence we employ a thresholding technique which divides the collection into those considered to contain the concept and those which do not. To achieve this, while simultaneously selecting a threshold value for each concept, we use the Kapur automatic thresholding technique [18]. Since this entropy based non-parametric method does not require any training, it can be easily applied to such a broad collection. We consider any images above the threshold value to be positive examples of that concept. Similarly, any frames below the threshold were considered as negative. Next, nine participants manually judged a subset of system positive and negative examples for each concept. In order to judge the intercoder reliability, consistency and accuracy of each annotator's performance; 50 positive and 50 negative examples per concept were randomly selected for judgment by each of the 9 annotators. Additionally, per concept, another 150 system judged positive and negative frames were randomly selected and assigned to every annotator. This resulted in almost 1400 positive and negative unique images per concept to be judged by the 9 annotators (50 to be judged by all 9 plus 9x150 individual judgments).

To support this judgment process a custom annotation tool was developed. Participants were presented with a tiled list of images and given instructions on how to

Table 4. Accuracy of detection for each concept (Sorted by 'System Positive Accuracy')

Concept Name	No. Samples Provided	Number of Judgements	System Positive Accuracy	System Negative Accuracy
Indoor	19,287	3,271	82%	45%
Sky	1,513	4,099	79%	90%
Screen	10,601	3,761	78%	85%
Shopping	228	3,500	75%	99%
Office	7,783	3,436	72%	77%
steeringWheel	208	3,936	72%	99%
Door	579	3,512	69%	86%
Hands	10,649	3,399	68%	68%
Veg	778	3,336	64%	97%
Tree	957	3,736	63%	98%
Outdoor	3,292	3,807	62%	97%
Face	3,018	3,452	61%	91%
Grass	444	3,765	61%	99%
insideVehicle	2,226	3,604	60%	93%
Buildings	1,853	3,654	59%	98%
Reading	927	3,420	58%	94%
Toilet	174	3,683	58%	99%
Stairs	87	2,927	48%	100%
Road	1,008	3,548	47%	96%
vehiclesExternal	480	3,851	46%	98%
People	6,633	3,024	45%	90%
Eating	2,143	3,530	41%	97%
holdingPhone	233	3,570	39%	99%
holdingCup	418	3,605	35%	99%
Meeting	2,396	3,534	34%	94%
presentation	352	3779	29%	99%
viewHorizon	3	3168	23%	98%

appropriately judge them against each concept. Users simply clicked an image to mark it as a positive match to the provided concept. For each concept both system judged positive and negative images were presented in tandem and were randomly selected from the total pool of judgments to be made. Annotating in this fashion allowed a total of 95,907 judgments made across all users on 70,659 unique concept validation judgments (which used 58,785 unique images). This yielded a detailed validation of both the images considered positive and negative for each concept.

Each annotator provided judgments for a shared set of 100 images per concept. These images were then used to determine the amount of agreement in the judgments among the nine annotators. An understanding of this 'intercoder reliability' is important as it validates the reliability of the overall annotation process and the performance of the annotators in general. This allows us to ensure that the outcome of the validation process is wholly reliable. The intercoder reliability was determined to be 0.68 for all judgments completed using Fleiss's Kappa [9]. As such the annotations provided by these participants are consistent and have very good inter-coder agreement. Examination at the concept level shows 18 of the 27 concepts had at least 0.6 agreement which is substantial according to Landis and Koch [19]. While examination of individual concepts reveals some variability in inter-rater reliability and a number of lower than anticipated agreement for a minority of the concepts (k=0.64 average overall; minimum 0.37 – view of horizon; maximum 0.86 – steering wheel), given that the number of judgments made per annotator was extremely large, this may have had the affect of reducing the overall magnitude of the value. We believe that the agreement between the annotators is sufficiently reliable to use these judgments to validate the automatically detected concepts.

3.1 Analysis of System Results

From the 95,907 judged results, 72,143 (75%) were determined to be correctly classified by the system. This figure, however, includes both positive and negative images for a concept as determined by the system. Of all those judgments, the system correctly identified 57% of true positives overall. 93% of system negatives were correct, meaning that only 7% of true positives were missed across all the concepts on the entire dataset.

Given the variation in complexity of the concepts and in the level of semantic knowledge they attempt to extract, it is unsurprising that there is notable differences in their performance and accuracy. Furthermore, the quality, variance and number of training examples will impact on the performance of an individual detector and as such these may be factors in their differing performances. This is outlined in Table 4 which is ordered by concept performance. From this it is clear that the 'indoor' detector worked best, with several other concepts providing similarly high degrees of accuracy. These include the "steeringWheel", "office", "shopping", and "screen" concepts. It is also interesting to note from Table 4, that with the exception of the 'indoor' concept, there are very few missed true positive examples in our large set of judged images. As the images were collected from 5 separate users it is interesting to explore the degree of variance in the performance between concepts (in terms of true positives). The performance ranged from 46% to 72%, but as illustrated in Fig 3, the deviation of results is not so large when the number of concept training samples

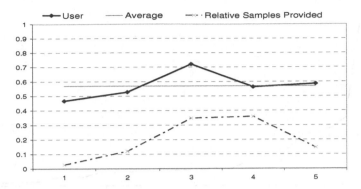

Fig. 3. Performance of all concepts on users' collections

provided to the system is considered (the blue dashed line at the bottom of Figure 3). There exists a strong correlation of 0.75 between the number of examples provided by each user to the system and the actual system classification results on the set of 95,907 judged results.

17 of the 27 concepts are at least 58% accurate in correctly identifying positive image examples for a given concept. Apart from the "people" concept we argue that the performance of the other concepts can be improved by providing more positive labeled image examples for each concept. We believe the concept detection results on SenseCam images are sufficiently reliable such that inferences on user patterns and behaviour may be made.

4 Event-Based Results of Concept Detection Activity

The concept detection results provided for the 219,312 images across the five content owners were then further analysed and investigated. There was wide variance in the number of images determined to be relevant from concept to concept. For example, 107,555 "indoor" images were detected, while just 72 images were detected as being of the "viewHorizon" concept. A number of concepts have a semantic relationship. For example, the "tree" (5,438) and "vegetation" (4,242) concepts closely relate to one another and as such have a similar number of positive examples. However, conversely, within the collection there were many more images containing "people" (29,794) than "face" (11,516) concepts. This is initially a little counterintuitive. While this may be attributed to the 'people' detector being relatively unreliable, there is another more probable explanation. Often a wearer will be for example on the street walking, or on a bus, and faces will not be clearly identifiable either as a result of people facing away from the wearer or being in the distance.

All of the collection owners are researchers and it was also noticed that the concepts with the highest number of occurrences closely match that of what would be expected for such users, e.g. "indoor", "screen", "hands" (e.g. on keyboard), "office", "meeting", etc. It should be noted that the concept detectors returns results that quite accurately fit those concepts that our users most commonly encounter.

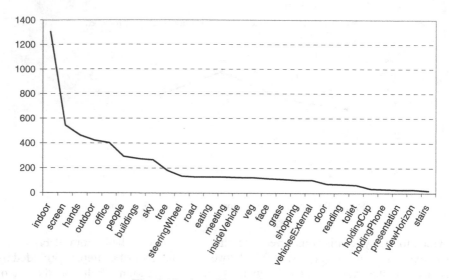

Fig. 4. Number of "events" per concept

As previously mentioned, using techniques as outlined in [8], a collection of SenseCam images can be aggregated into a higher level discrete unit known as an 'event'. This has the function of reducing the approximate 3,000 images captured in an average day to on average 20 'events', making the collection far more manageable for its owner. It also has the added advantage of more closely approximating the 'episodic' units in which we as humans consider our past experiences. While our analysis has focused on the accuracy of the concept detectors at the image level, it is, as such, worth considering the concepts as detected at the higher level of 'events'.

Using this event segmentation, we identified a total of 3,030 events in the collection. As a concept will likely be related to the higher-level activity embodied by an event, the concepts were further explored at the event level. To achieve this we determined the concentration of the concepts within each event, e.g. a 'working at the computer' event will have a high percentage of images containing the "screen" and "indoor" concept. To be consistent with our prior analysis on the image-based level, we again made use of the Kapur thresholding approach [18] to determine whether an event had a sufficiently dense concentration of images of a particular concept. That is to say, for each concept we determined the percentage of images in an event that must contain that concept, in order for that event to be considered to represent the concept. "Building events", for example, should have at least 28% of their images identified as being "building", while "indoor" events should have at least 48% of images with the "indoor" concept, etc. It is evident from observation of Fig. 4, the event and image level are very similar in their distribution of concepts. As expected the "indoor", "screen", and "office" concepts are still very common when they are considered in terms of events. Likewise, there is a very small sample of events that are under the concept types of "stairs", "viewHorizon", "holdingPhone", etc.

It is particularly interesting to consider, as illustrated by Fig. 5, the number of events each user had (relative to the size of their collection) for each concept type. To explain further, for user 1 the "steeringWheel" concept occurred over 350% more

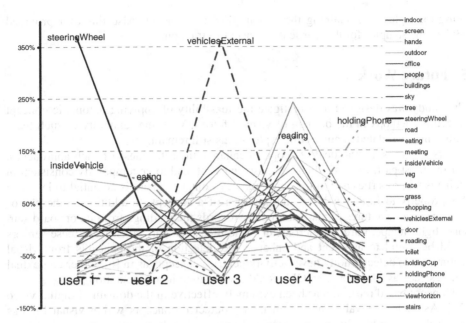

Fig. 5. Deviation of user examples to the median (per concept)

frequently than the median of all the other users. The median value is the x-axis, i.e. 0% different to the median! As such, this graph gives an outline of the differing life-styles of the users.

For user 1 it is interesting to observe that he has much more "steeringWheel" and "insideVehicle" events than the other users. This is indeed to be expected given that this user is the only one of our collectors who regularly drove a car. In fact in providing the initial set of 208 positive examples of this concept, user 1 was responsible for all of these images.

For user 2 it is noticeable that there are relatively many more "eating" events. An explanation for this is that this user wore the SenseCam to quite a few conferences, which included quite a few meals. Also this user was generally quite diligent in wearing his SenseCam for breakfast and supper. It is interesting to note that this user did not provide the most samples for this concept detector to train on initially.

For user 3 there were many more "vehiclesExternal" events than for the other users. We attribute this to the fact that this user provided 67% of the samples for the concept detector to train on. While with user number 4, it is quite evident that he has many more "reading" events. An explanation is that this researcher is very diligent in terms of reading his literature, and is well known for this trait.

User number 5 seemingly had an unusually high number of "holdingPhone" events. We explain this by the fact that this user was conducting experiments with his mobile phone at the time of capturing these SenseCam images. Due to the nature of the experiment he was additionally capturing surrogate image data using the mobile phone's camera and as such was carrying the phone throughout the data collection period. As a result many of his events (relative to the other users) were annotated as being examples of containing the "holdingPhone" concept. Also this user provided 63% of the samples for this particular concept to train on.

5 Future Work

This study was designed to investigate the feasibility of applying automatic concept detection methods in the domain of visual lifelogs. With the reliability of such techniques now validated, a number of explorations still remain.

First, there is a great deal of scope to enhance the robustness of such approaches. For the most part, frames which compose an event tend to be temporally consistent in their visual properties and in the concepts they contain. There is potential to leverage this property to further validate the presence of a concept. In addition to the photos the SenseCam captures, it also continually records the readings from its onboard sensors (light, temperature, accelerometer). The measurements taken from these sensors could be useful to augment and enhance the detection of the concepts from visual features or to detect wholly new 'activity-centric' concepts as in [6]. Other contextual sources such as Bluetooth and GPS could also be used in augmentation [4].

Concept based retrieval has been extremely effective in the domain of digital video [24]. As such retrieval using automatically detected concepts within visual lifelogs should be explored. The performance and utility of such concept-based retrieval approaches should be compared with other methods such as using social context [4].

Finally, exploration into semi-automatic concept annotation of a collection could be achieved through active learning. This would offer the ability to create and train new concept detectors as users explore and annotate their collections. By enabling efficient automatic annotation of new content, while providing flexibility to users to personalise and extend the set of concepts, further utility would be added to lifelogs.

6 Conclusions

In order to fulfill Bush's MEMEX vision we must seek to offer rapid and flexible access to the contents of a visual, multimedia lifelog. However, as such collections are extremely voluminous and ever-growing, this is particularly challenging. Manual browsing or annotation of the collection to enable retrieval is impractical and we must seek automatic methods to provide reliable annotations to the contents of a visual lifelog. We have documented the process of applying automatic detectors for 27 everyday semantic concepts to a large collection of SenseCam images, and rigorously validated the outcomes. Nine annotators manually judged the accuracy of the output for these 27 concepts on a subset of 95,000 lifelog images spanning five users. We found that while the concepts' accuracy is varied, depending on the complexity and level of semantics the detector tried to extract from an image, they are largely reliable and offer on average a precision of 57% for positive matches and 93% for negative matches within such a collection.

Furthermore, using the output of the concept detection process, we have been able to identify trends and make inferences into the lifestyles of our 5 users. These inferences were based on the system judgments made for the 27 concepts on an extended collection of almost 220,000 images. By intelligently correlating semantic concepts with previously segmented events or 'activities', we have been able to determine the occurrence of a concept in the users' activities e.g. user 2 has 52 eating events. We have determined through qualitative means that this approach is promising for the identification of concept patterns which occur within an individual's visual lifelog and more generally the concepts of interest and importance for an individual.

These results are particularly encouraging and suggest that automatic concept detection methods translate well to the novel domain of visual lifelogs. Once applied to such a collection it offers the ability to enable a range of opportunities, with the most important being the efficient automatic annotation and retrieval within such a voluminous collection.

Acknowledgements

We are grateful to the AceMedia project and Microsoft Research for support. This work is supported by the Irish Research Council for Science Engineering and Technology, by Science Foundation Ireland under grant 07/CE/I1147 and by the EU IST-CHORUS project. We also would like to extend our thanks to the participants who made their personal lifelog collection available for these experiments, and who partook in the annotation effort.

References

1. Bell, G., Gemmell, J.: A Digital Life. Scientific American (2007)
2. Bovik, A.C., Clark, M., Geisler, W.S.: Multichannel texture analysis using localized spatial filters. IEEE Transactions on Pattern Analysis and Machine Intelligence 12(1), 55–73 (1990)
3. Bush, V.: As We May Think. Atlantic Monthly 176(1), 101–108 (1945)
4. Byrne, D., Lavelle, B., Doherty, A.R., Jones, G.J.F., Smeaton, A.F.: Using Bluetooth and GPS Metadata to Measure Event Similarity in SenseCam Images. In: Proc. of IMAI 2007 (July 2007)
5. Chang, C.C., Lin, C.J.: LIBSVM: a library for support vector machines (2001), http://www.csie.ntu.edu.tw/~cjlin/libsvm/
6. DeVaul, R.W., Dunn, S.: Real-Time Motion Classification for Wearable Computing Applications, Project paper (2001), http://www.media.mit.edu/wearables/mithril/realtime.pdf
7. Doherty, A.R., Byrne, D., Smeaton, A.F., Jones, G.J.F., Hughes, M.: Investigating Keyframe Selection Methods in the Novel Domain of Passively Captured Visual Lifelogs. In: Proc. of the ACM CIVR 2008, Niagara Falls, Canada (2008)
8. Doherty, A.R., Smeaton, A.F.: Automatically Segmenting Lifelog Data Into Events. In: Proc. 9th International Workshop on Image Analysis for Multimedia Interactive Services (2008)
9. Fleiss, J.L.: Measuring nominal scale agreement among many raters. Psychological Bulletin 76(5), 378–382 (1971)
10. Fuller, M., Kelly, L., Jones, G.J.F.: Applying Contextual Memory Cues for Retrieval from Personal Information Archives. In: Proceedings of PIM 2008 Workshop Florence, Italy (2008)

11. Geusebroek, J.M.: Compact object descriptors from local colour invariant histograms. In: British Machine Vision Conference, Edinburgh, UK (2006)
12. Geusebroek, J., Smeulders, A.W.M.: A six-stimulus theory for stochastic texture. International Journal of Computer Vision 62(1/2), 7–16 (2005)
13. Gurrin, C., Smeaton, A.F., Byrne, D., O'Hare, N., Jones, G.J., O'Connor, N.: An examination of a large visual lifelog. In: Li, H., Liu, T., Ma, W.-Y., Sakai, T., Wong, K.-F., Zhou, G. (eds.) AIRS 2008. LNCS, vol. 4993, pp. 537–542. Springer, Heidelberg (2008)
14. Hoang, M., Geusebroek, J., Smeulders, A.W.M.: Color texture measurement and segmentation. Signal Processing 85(2), 265–275 (2005)
15. Hodges, S., Williams, L., Berry, E., Izadi, S., Srinivasan, J., Butler, A., Smyth, G., Kapur, N., Wood, K.: SenseCam: A Retrospective Memory Aid. In: 8th International Conference on Ubiquitous Computing, Orange County, USA, pp. 177–193 (2006)
16. Jiang, Y.G., Ngo, C.W., Yang, J.: Towards Optimal bag-of-features for Object Categorization and Semantic Video Retrieval. In: Proceedings of the ACM International Conference on Image and Video Retrieval, Amsterdam, The Netherlands, pp. 494–501 (2007)
17. Jurie, F., Triggs, B.: Creating efficient codebooks for visual recognition. In: IEEE International Conference on Computer Vision, Beijing, China, pp. 604–610 (2005)
18. Kapur, J.N., Sahoo, P.K., Wong, A.K.C.: A New Method for Graylevel Picture Thresholding using the Entropy of the Histogram. Comp. Vis., Grap., & Image Proc. (1985)
19. Landis, J.R., Koch, G.G.: The measurement of observer agreement for categorical data. Biometrics 33, 159–174 (1977)
20. Lee, H., Smeaton, A.F., O'Connor, N.E., Jones, G.F.J.: Adaptive Visual Summary of LifeLog Photos for Personal Information Management. In: Proc. 1st Intnl. Workshop on Adaptive Infor. Retrieval, pp. 22–23 (2006)
21. Lin, H.T., Lin, C.J., Weng, R.: A note on Platt's probabilistic outputs for support vector machines. Machine Learning 68(3), 267–276 (2007)
22. Naphade, M.R., Kennedy, L., Kender, J.R., Chang, S., Smith, J.R., Over, P., Hauptmann, A.: A Light Scale Concept Ontology for Multimedia Understanding for TRECVID 2005. Technical Report RC23612, IBM T.J. Watson Research Center (2005)
23. Snoek, C.G.M., Worring, M., van Gemert, J.C., Geusebroek, J.M., Smeulders, A.W.M.: The Challenge Problem for Automated Detection of 101 Semantic Concepts in Multimedia. In: ACM Multimedia 2006, Santa Barbara, USA, pp. 421–430 (2006)
24. Snoek, C.G.M., van Gemert, J.C., Gevers, T., Huurnink, B., Koelma, D.C., van Liempt, M., de Rooij, O., van de Sande, K.E.A., Seinstra, F.J., Smeulders, A.W.M., Thean, A.H.C., Veenman, C.J., Worring, M.: The MediaMill TRECVID 2006 semantic video search engine. In: Proceedings of the 4th TRECVIDWorkshop, Gaithersburg, USA (2006)
25. van Gemert, J.C., Snoek, C.G.M., Veenman, C.J., Smeulders, A.W.M., Geusebroek, J.M.: Comparing compact codebooks for visual categorization. Computer Vision and Image Understanding (submitted, 2008)
26. Vapnik, V.: The Nature of Statistical Learning Theory, 2nd edn. Springer, New York (2000)
27. Wang, D., Liu, X., Luo, L., Li, J., Zhang, B.: Video Diver: generic video indexing with diverse features. In: Proceedings of the ACM SIGMM International Workshop on Multimedia Information Retrieval, Augsburg, Germany, pp. 61–70 (2007)
28. Yanagawa, A., Chang, S.F., Kennedy, L., Hsu, W.: Columbia university's baseline detectors for 374 LSCOM semantic visual concepts. Technical Report 222-2006-8, Columbia University ADVENT Technical Report (2007)

Using Fuzzy DLs to Enhance Semantic Image Analysis

S. Dasiopoulou[1,2], I. Kompatsiaris[2], and M.G. Strintzis[1]

[1] Information Processing Laboratory, Electrical and Computer Engineering
Department, Aristotle University of Thessaloniki, Greece
[2] Multimedia Knowledge Laboratory, Informatics and Telematics Institute,
Thessaloniki, Greece

Abstract. Research in image analysis has reached a point where detectors can be learned in a generic fashion for a significant number of
conceptual entities. The obtained performance however exhibits versatile
behaviour, reflecting implications over the training set selection, similarities in visual manifestations of distinct conceptual entities, and appearance variations of the conceptual entities. In this paper, we investigate
the use of formal semantics in order to benefit from the logical associations between the conceptual entities, and thereby alleviate part of the
challenges involved in extracting semantic descriptions. More specifically,
a fuzzy DL based reasoning framework is proposed for the extraction
of enhanced image descriptions based on an initial set of graded annotations, generated through generic image analysis techniques. Under
the proposed reasoning framework, the initial descriptions are integrated
and further enriched at a semantic level, while additionally inconsistencies emanating from conflicting descriptions are resolved. Experimentation in the domain of outdoor images has shown very promising results,
demonstrating the added value in terms of accuracy and completeness
of the resulting content descriptions.

1 Introduction

Digital image content is omnipresent on the Web; Google posted on August 2005,
a total image size of $2,187,212,422$, Yahoo estimated that its index covered
1.5 billion of images at that time, while nowadays statistics show a continuous
growth in these numbers (indicatively Flickr uploads amount to an average of
about 3000 images per minute). Given such numbers, the availability of machine
processable semantic descriptions for this content becomes a key factor for the
realisation of applications of practical interest, perpetuating the challenge of
what constitutes the multimedia community holy grail, i.e. the *semantic gap*
between representations that can be automatically extracted and the underlying
meaning [1].

In the late 1970s and early 1980s, influenced by the Artificial Intelligence (AI)
paradigm, image analysis and understanding became a problem of acquiring intelligent behaviour through computational means, resulting in the first attempts

D. Duke et al. (Eds.): SAMT 2008, LNCS 5392, pp. 31–46, 2008.

towards knowledge-directed image analysis. A period of explosive growth in approaches conditioned by knowledge followed [2,3]: varying knowledge representation and reasoning schemes, in accordance with the contemporary AI assets, were proposed, and knowledge attempted to address all aspects involved, ranging from perceptual characteristics of the visual manifestations to control strategies. The broad and ambitious scope targeted by the use of knowledge, resulted in representations and reasoning mechanisms that exhibited high complexity and inflexibility. As a result, research shifted to modular architectures, treating separately the individual subproblems. Machine learning (ML) approaches gained increased popularity as means to compensate for the complexity related to the explicit representation of associations between perceptual features and conceptual entities. Treating however, semantics solely as embodiments of visual manifestations, bears a number of serious limitations: not all of semantics can be expressed solely with visual means, variations pertain the possible manifestations of a single semantic entity, and distinct semantic entities share similar manifestations. Adding to the aforementioned the implications of the choices related to the training process, machine learning approaches, despite having reported satisfactory performances for given datasets, tend to scale poorly as the number of considered concepts is increased or when new content is introduced.

The advent of the Semantic Web paved a new era in knowledge sharing, reuse and interoperability, by making formal semantics explicit and accessible to heterogenous agents and applications. The image analysis community embraced the new technologies, utilising ontologies at first in order to attach meaning to the produced annotations, and subsequently as means for assisting the very extraction of the annotations [4,5,6,7,8]. A characteristic shared among these approaches is that uncertainty is poorly handled. In the approaches addressing the transition from perceptual features to conceptual entities, thresholds and allowed ranges regarding the values of the considered features are used, i.e. ambiguity is treated as a separate aspect from the domain semantics [5,4,9,10]. On the other hand, in the approaches that focus more on the utilisation of semantics and inference for the purpose of acquiring descriptions of higher complexity, uncertainty is not taken into consideration at all [8,11].

Acknowledging the inherent role of uncertainty in image understanding, in this paper we propose a formal reasoning framework for the extraction of semantically coherent image descriptions, extending the earlier work presented in [12]. More specifically, we investigate the recent advances in fuzzy Description Logics (DLs) extensions [13,14], under a generic analysis context with the purpose of overcoming part of the limitations that result from the non semantic view taken by machine learning techniques. The input of the proposed reasoning framework consists of ML obtained descriptions, along with the calculated distance values, i.e. the values indicating the similarity from the training set concepts. The graded descriptions are treated as fuzzy assertions, where each concept corresponds to a fuzzy set specified by the values learned for the parameters comprising the internal ML structure, and the corresponding distance reflects the membership of the examined image (image segment) to a given concept. The ML based extracted descriptions may refer to different levels of granularity, and no assumptions are made regarding specific implementation issues. The possibly overlapping, contradictory or complementary, input fuzzy assertions are integrated and interpreted through the utilisation of formal semantics and reasoning.

The rest of the paper is organised as follows. Section 2 discusses the task of semantic image analysis, indicating issues and requirements related to the utilisation of formal semantics and reasoning. Section 3 introduces fuzzy DLs from an image analysis perspective, exemplifying their usability and potential. In Section 4, we present the implementation details of the proposed reasoning framework, and in Section 5, the results of the evaluation. Section 6 reports on relevant work, and finally Section 7 concludes the paper.

2   Applying Reasoning in Semantic Image Analysis

The use of formal, explicit knowledge, and subsequently inference, in semantic image analysis aims to exploit background knowledge that is characteristic of the semantic entities (objects, events) of interest within a particular context. The expected value from formally representing the semantics of a domain can be roughly summarised as "assisting the extraction of descriptions by making explicit the way semantics are perceived, ensuring thus the acquisition of interpretations that match human cognition". In practical terms, the use of knowledge translates into the elimination of semantic inconsistencies and the acquisition of descriptions whose interpretation goes beyond visual patterns.

Under these considerations, using formal languages to represent mappings between feature values and domain entities, such as in [4], where colour, texture and shape values are mapped to tumour types, or [10], where colour and shape values are mapped to natural objects, may be significant for purposes of sharing and reusing knowledge, but it does not leave much opportunities for utilising reasoning in terms of intelligence through computational means. It is not simply a matter of the limited datatype support provided by ontology languages such as RDFS[1] and OWL[2], but because of the non logical nature of the problem at hand, i.e. the estimation of distance between a given data structure that constitutes a feature model and the measurable feature values. As the image and knowledge representation communities get more familiar to each other, the role and potential of formal knowledge inference, in semantic image analysis is revised and reassessed. In the following we discuss common issues that arise in image analysis and how they relate to the application of reasoning.

2.1   Issues and Requirements

Still image analysis, depending on the application context, may refer to the extraction of descriptions at segment level, image level, or both. In the first case, the extracted descriptions refer to notions corresponding to objects and

---

[1] http://www.w3.org/TR/rdf-schema/
[2] http://www.w3.org/TR/owl-features/

are accompanied by localisation information. In the second case, the descriptions apart from global scene characteristics may as well indicate the presence of objects, however without providing any information about their location. In either case, low-level features are correspondingly extracted and processed, so that the associations that underly features and conceptual entities can be captured. Independently from the learning approach followed, the extracted associations reflect correspondences between visual manifestations and conceptual notions. These correspondences are neither sound, due to the similarities in appearance between distinct semantic entities, neither complete, due to the inability to capture all possible manifestations a semantic entity may have. Furthermore, as they rely solely on visual characteristics, their potential addresses strictly meaning that can be related to visual information. The aforementioned result in a number of implications which provide the context for the identification of requirements with respect to the utilisation of formal semantics and reasoning.

A first requirement refers to the ability to handle contradictory descriptions at the same level of granularity, e.g. at image (scene) or segment level. In the case of scene level descriptions, it is not adequate to choose the description with the highest plausibility, where plausibility may refer to the probability of the description being true or to the degree to which this description is true. The various scene level descriptions are usually logically related to each other, the simplest case being that of addressing descriptions of increasing detail. Consequently, the plausibility of each extracted description is intertwined to the plausibility of descriptions referring to the semantically related notions. Similar observations hold for the case of segment level descriptions, when the extracted descriptions address only the presence of concepts. When localisation information is also available, image processing issues arise as well, since apart from semantic coherency, the extracted descriptions need to be checked at the partitioning plane (or planes, as different implementations may adopt independent segmentations).

A second requirement refers to the ability to handle and resolve semantic conflicts between descriptions of different granularity. As in the previous case, the conceptual notions at object and scene levels are highly likely to be logically related. The visual information that the learning approaches rely on, cannot ensure that such relations are reflected in the learned associations. Such limitation applies to both directions. The plausibility of object level concepts may impact on the existence and plausibility of scene level concepts, e.g. the presence of mountain indicates high plausibility for a mountainous scene, and conversely. Similarly, the presence of a given scene level concept may affect the plausibility of a concept at object level. For example, a scene description of beach affects (negatively) an object level description of snow.

A third requirement involves the ability to support descriptions, at scene or segment level, whose semantics lie beyond the potential of visual characteristics, i.e. semantics that originate in logical associations between other concepts. Examples include cases of part-whole relations, temporal sequences of simpler concepts, the combined or exclusive presence of given concepts, etc. Contrary to the aforementioned, this requirement impacts mostly on the selection of the knowledge representation scheme, as it embodies the main goal of knowledge representation and reasoning, i.e. the automatic extraction of implicit knowledge based on the explicitly available one. As in every knowledge-based application, the construction of the knowledge depends on the application context. For example, the concept of swimmer is more likely to be described as a face surrounded by sea, than a person in the sea.

Underlying the aforementioned requirements is the need to provide the means for handling uncertainty. This forms a crucial requirement, due to the inherent in image analysis ambiguity, where visual information plays the role of cues rather than evidences, allowing the detection of a concept only with an estimated plausibility. It is important to note that the semantics of the ambiguity are not unique. When analysis follows a similarity based approach, matching extracted feature values to prototypical ones, the resulting plausibility represents the distance from the prototypical values, i.e. the degree to which the examined image (segment) belongs to a given concept. On the other hand, when a Bayesian approach has been followed, the calculated values represent a different kind of ambiguity, i.e. the probability that the examined image (segment) is an instance of the examined concept. Naturally, under an analysis framework both kinds of ambiguity may be present.

In the following we present a brief introduction to the basics of fuzzy DLs and discuss how the aforementioned requirements relate to core DL inference services. Additionally, we indicate the functionalities, that once available would allow the extraction of image semantics under a fuzzy DL based reasoning framework. We note that in the following, uncertainty refers strictly to cases of machine learning based analysis, i.e. to fuzzy logic semantics. A combined fuzzy-probabilistic reasoning framework, although very interesting and promising towards a more complete confrontation of the image understanding challenge, is currently beyond the scope of this paper.

3   Fuzzy DLs and Semantic Image Analysis

Description Logics (DLs) [15] are a family of knowledge representation formalisms characterised by logically founded formal semantics and well-defined inference services. Starting from the basic notions of atomic concepts and atomic roles, arbitrary complex concepts can be described through the application of corresponding constructors (e.g., $\neg$, $\sqcap$, $\forall$). Terminological axioms (TBox) allow to capture equivalence and subsumption semantics between concepts and relations, while real world entities are modelled through concept $(a : C)$ and role $(R(a, b))$ assertions (ABox). The semantics of DLs are formally defined through an interpretation I. An interpretation consists of an non-empty set Δ^I (the domain of interpretation) and an interpretation function $.^I$, which assigns to every atomic concept A a set $A^I \subseteq \Delta^I$ and to every atomic role R a binary relation $R^I \subseteq \Delta^I x \Delta^I$. The interpretation of complex concepts follows inductively [15].

In addition to the means for representing knowledge about concepts and assertions, DLs come with a powerful set of inference services that make explicit

Table 1. Example of TBox

Natural ≡ Outdoor ⊔ ¬ ManMade
∃ contains.Sky ⊑ Outdoor
Beach ≡ ∃contains.Sea ⊓ ∃contains.Sand
Beach ⊑ Natural
Cityscape ≡ ∃contains.(Road ⊔ Car) ⊔ ∃contains.Building
Cityscape ⊑ ManMade

the knowledge implicit in the TBox and ABox. *Satisfiability, subsumption, equivalence* and *disjointness* constitute the main TBox inferences. Satisfiability allows to check for concepts that correspond to the empty set, subsumption and equivalence check whether a concept is more specific or respectively identical to another, while disjointness refers to concepts whose conjunction is the empty set. Regarding the ABox, the main inferences are *consistency*, which checks whether there exists a model that satisfies the given knowledge base, and *entailment*, which checks whether an assertion ensues for a given knowledge base.

Assuming a TBox that describes a specific domain, one can build an ABox from the analysis extracted descriptions and benefit from the inferences provided to detect inconsistencies and obtain more complete descriptions. Let us assume the TBox of Table 1, and the analysis extracted assertions *image1* :*Cityscape*), *region1* :*Sea*), *region2* :*Sand*), (*image1,region1*) : *contains*, and (*image1,region1*) : *contains*. The following assertions entail: (*image1* :*ManMade*), (*image1* :*Beach*), and (*image1* :*Natural*). Furthermore an inconsistency is detected, caused by the disjointness axiom relating the concepts Natural and ManMade. However, the aforementioned apply only in the case of crisp assertions, which is not the common case in image analysis. Applying thresholds in the initial descriptions and transforming them to binary, does not overcome the problem; instead additional issues are introduced. Let us assume, the following set of initial descriptions, (*region1* : *Sea*) ≥ 0.8, (*region1* : *Sand*) ≥ 0.9, (*image1* : *Cityscape*) ≥ 0.9, and that all role assertions involving the role *contains* have a degree of ≥ 1.0. Transforming them directly to crisp, would result in the crisp assertions (*region1* : *Sea*), (*region1* : *Sand*), and (*image1* : *Cityscape*), which would cause an inconsistency with no clear clues about which assertions are more prevailing, as the degree information has been omitted.

In the case of a fuzzy DL language, the ABox consists of a finite set of *fuzzy assertions* of the form $a : C \bowtie n$ and $(a,b) : R \bowtie n$, where $\bowtie$ stands for ≥, >, ≤, and <[3]. The semantics are provided by a fuzzy interpretation, which in accordance to the crisp DLs case, is a pair $I = (\Delta^I, \cdot^I)$ where Δ^I is a non-empty set of objects called the domain of interpretation, and $\cdot^I$ is a *fuzzy*, this time, interpretation function which maps: an individual a to an element $a^I \in \Delta^I$, i.e., as in the crisp case, a concept name A to a membership function $A^I : \Delta^I \to [0,1]$, and a role name R to a membership function $R^I : \Delta^I \times \Delta^I \to [0,1]$.

---

[3] Intuitively a fuzzy assertion of the form $a : C \geq n$ means that the membership degree of the individual a to the concept C is at least equal to n.

Regarding fuzzy DLs extensions, two main efforts exist currently that address formally both the semantics and the corresponding reasoning algorithms. In [14,16], the DL language *SHIN* has been extended according to fuzzy set theory leading to the so called f-*SHIN*. The fuzzy extensions address the assertion of individuals and the extension of the language semantics. In [13], a fuzzy extension of *SHOIN(D)* is presented, which constitutes a continuation of earlier works of the authors on extending *ALC*, *SHIF*, and *SHIF(D)* to fuzzy versions [17,18]. In addition to extending the *SHOIN(D)* semantics to f-*SHOIN(D)*, the authors present a set of interesting features: concrete domains as fuzzy sets, fuzzy modifiers such as *very* and *slightly*, and fuzziness in entailment and subsumption relations.

Additionally to the theoretic foundations for the fuzzy extensions, respective reasoning algorithms have been presented and implemented, namely the *Fuzzy Reasoning Engine* (FiRE)[4] and the *fuzzyDL*[5].

Continuing the previous example, let us assume that the following fuzzy assertions result from analysis: $(image1 : Cityscape) \geq 0.4$, $(image1 : Outdoor) \geq 0.82$, $(region1 : Sea) \geq 0.8$, $(region2 : Sand) \geq 0.65$, $(region3 : Sky) \geq 0.9$, and $(image1, region_i) : contains \geq 1.0$, for $i = 1, 2, 3$. Querying for the greatest lower bound for the individual *image1* with respect to the concepts Outdoor and Beach we retrieve 0.9 and 0.65 respectively. Note that Beach is defined as the conjunction of two existential restrictions involving the fillers Sea and Sand. Since the assertions referring to the role contains have a degree ≥ 1.0, the degrees of the two existential definitions depends only on the degrees of their respective fillers. Under Zadeh's semantics, the T-norm equals the minimum of the involved degrees, i.e. $min\{0.65, 0.8\} = 0.65$. The $\exists$ contains.Sky $\sqsubseteq$ Outdoor axiom, gives greatest lower bound for $(image1 : Outdoor)$ equal to 0.9, updating the explicitly given value of 0.82. As the detection of an outdoor image does not give any further information about a more specific scene description or of the objects that may be depicted, the degrees of the other assertions should not be affected. On the contrary the presence of any of the three concepts, means that it is an outdoor concept, with a plausibility greater or equal to the corresponding degree of the respective concept. The subsumption axioms between the concepts Beach, Natural and Cityscape capture this knowledge, ensuring an appropriate behaviour.

Furthermore, let us examine in a greater detail the implication of the fuzzy conjunction semantics appearing in axiom Beach $\equiv$ $\exists$contains.Sea $\sqcap$ $\exists$contains.Sand. Under a case where for one of the existential restriction fillers there can be no assertion, no entailment can be made with respect to the Beach concept. Assuming now an assertion of the Beach concept, it entails the presence of assertions for the Sea and Sand concepts, and with degrees greater or equal to that of the Beach assertion. One can observe that again the fuzzy DLs semantics reflect the desired behaviour. Note that in the fuzzy case, disjointness has different semantics than in the crisp case, i.e. two concepts that are disjoint

---

[4] http://www.image.ece.ntua.gr/~nsimou
[5] http://faure.isti.cnr.it/~straccia/software/fuzzyDL/fuzzyDL.html

raise no inconsistency as long as the T-norm (conjunction) of their degrees is not equal or greater than 0.5. Generally, in the examined image analysis context, a degree less than 0.5 indicates a rather poor match in terms of visual similarity, resulting in practically ignoring the corresponding assertions. We note however, that in case the corresponding concepts classifiers are implemented in a binary fashion, i.e. low values indicate the non presence of a concept, then such fuzzy assertions play an important role, as they entail assertions of high degrees for the negated concepts.

We observe that fuzzy DLs, through the expressivity and the inference services that they proffer, constitute a very promising technology for supporting the extraction of image semantics. The available functionalities allow one to formally describe the different kinds of logical associations that define the semantics of the concepts and roles of the respective domain of interest, and ensure the entailment of intended descriptions, the semantic based update of the degrees, and the detection of inconsistencies, in case the initial description are violating the semantic model of the domain. However, in order to utilise fuzzy DLs in semantic image analysis, there are still two main issues: the definition of a framework under which the requirements described in section 2.1 can be realised based on fuzzy DLs inference services, and the handling of inconsistencies, so that a final set of consistent descriptions can be obtained. Handling inconsistencies in DLs knowledge bases usually refers to approaches targeting revision of the terminological axioms [19,20]. In the examined case however, the inconsistencies result from the limitations in associating semantics with visual features. Thus it is the ABox that needs to be appropriately managed. The methodology that we followed, is described in the following section, where the implementation of the proposed fuzzy DLs based reasoning framework is detailed.

4   A Fuzzy DLs-Based Reasoning Framework for Semantic Image Analysis

In the previous Section we highlighted how fuzzy DLs relate to the tasks involved in supporting semantic image analysis. In this section, we present the details of the proposed fuzzy DLs based reasoning framework, which utilising the core fuzzy DLs inference services, accomplishes the requirements highlighted in Section 2.1. Summarising, the latter address three key issues: i) consistency checking and handling of assertions of the same granularity (i.e. scene and object level), ii) consistency checking and handling between assertions of different granularity, and iii) enrichment of the descriptions by means of logical entailment. In the current study, localisation information of object level descriptions is not taken into account, reducing the first task to consistency checking at scene level. Using the fuzzyDL system for the core fuzzy DLs inferences, the proposed reasoning framework realises the extraction of image semantics as follows. First, the descriptions that apply to an image at scene level are determined through reasoning, utilising the terminological box semantics. In the sequel, based on the previously inferred scene level descriptions, inconsistencies in the initial set

of scene and object level assertions are tracked and resolved, leading to a semantically meaningful description for the image. The last step refers to the en-
hancement of the description by making explicit assertions that result by logical
entailment. The details of each step are presented follow.

Selection of scene descriptions. Due to the logical associations between concepts
that refer to objects and concepts that refer to scene level notions, all the available analysis produced assertions need to be taken into account at this step.
First, the hierarchy of the scene level concepts is computed, based on the respective TBox. Starting from the more specialised scene concepts, i.e. concepts
that are not subsumed by other scene concepts, and moving upwards the hierarchy, the assertions of the corresponding concepts are processed with respect to
fuzzy semantics. In each level of the hierarchy, the assertion, explicit or inferred,
with the greatest degree prevails the assertions of disjoint concepts. The latter
are stored in a list as they indicate sources of inconsistency to be addressed in
the following step. In case of assertions referring to concepts that are not disjoint, all of them are preserved. Moving to the next level of the scene concepts
hierarchy, the procedure is repeated, checking additionally whether the prevailing concept of the current level is subsumed by the concept(s) selected at the
previous level. If a subsumption relation holds, the degree at the current level is
updated accordingly so that the degree of the subsumer is greatest or equal to
that of the subsumee.

In the opposite case, the concepts of the previous level are moved to the list of
inconsistent concepts. To give an example, assume that the concepts Beach and
Cityscape are at the same level of the hierarchy, and that the next level includes
the concepts Natural and ManMade to which the former are related through respective subsumption axioms. Assuming the assertions $(image1 : Beach) \geq 0.8$,
$(image1 : Cityscape) \geq 0.6$, and $(image1 : ManMade) \geq 0.9$ the Beach concept
is preserved at the first step. Moving to the next level however, the concept Man-
Made prevails that of Natural, which means that the Beach referring assertion
needs to be considered for inconsistency. The aforementioned process iterates
till reaching the top level concepts of the hierarchy. Obviously, the disjointness
axioms, need to be removed from the TBox on which fuzzy DLs reasoning is
performed, and handled separately, in order to prevent halting the inference in
case of contradictions.

Handling of inconsistency. The previous procedure results in the identification
of the concepts at scene level that are inferred as valid and of the scene level
concepts that constitute sources of inconsistency. The first step of the consistency handling tasks is to query for assertions that refer to object level concepts
that are directly disjoint to the previously selected scene level concepts. In the
existence of such assertions, if the referred concept is atomic, the assertions are
directly removed. If the referred concept is complex, the assertions that led to
its inference are tracked and based on the semantics of the DLs constructors
possible solutions are identified and stored. To give a simplified example, in the
case of a complex concept whose definition consists in the conjunction of atomic

concepts, the possible solutions equal the number of conjuncts. The next step, considers inconsistencies that arise due to disjointness axioms between scene level concepts, which includes the case of assertions referring to object level concepts that entail scene level assertions. Again, based on the terminological axioms the assertions that cause the inconsistency are tracked, and solutions are computed based on the involved constructors.

In order to enable the unhindered running of the fuzzy DL reasoner inference services, yet preserve all axioms in the TBox, based on the list of inconsistent concepts resulting during the first task, corresponding definitions of *non-concepts* are introduced, and respective subsumption axioms are added with respect to the original concepts. For example, assuming that Beach is included in the inconsistent concepts list, the axiom Beach⊑Non-Beach is added to the TBox, allowing thus the tracking of inconsistency without halting the reasoner. The alternative solutions calculated for resolving the inconsistencies are computed collectively, over the entire ABox, so that possible dependencies among them are taken into account. Checking and tracking all inconsistencies results in the general case in a set of possible solutions. In order to choose among the alternative solutions, we rank the set of solutions according to the number of assertions that need to be removed per solution, and the average of the corresponding degrees. The solution that involves the removal of the fewer assertions is eventually preferred (the average degrees are used to select between solutions of equal size).

Enrichment of descriptions. The enrichment of the descriptions by means of entailment is the most straightforward of the considered tasks. Once the scene level concepts are selected, and the assertions ensuing inconsistencies either directly or through complex definitions, are resolved, we end up with a semantically consistent set of assertions, that constitute the description of the image. Consequently, all that is left is to make explicit the assertions that are implicit in this final set. For this reason, appropriately queries are posed and the responses are included in the image description.

5   Experimental Results and Evaluation

In the previous sections, we described the reasons that motivated our investigation into a fuzzy DL based reasoning framework for supporting and enhancing semantics extraction from images. In order to assess the utilisation of formal semantics under the proposed reasoning framework, we carried out two experiments in the domain of outdoor images. An extract of the constructed Tbox is shown in Table 2. The test set consisted of 350 images, for which ground truth was manually generated at scene and object level according to the constructed TBox. In the evaluation, we compare the reliability of the image descriptions extracted through machine learning to that of the descriptions resulting after the application of reasoning. We adopted the precision, recall, and F-measure metrics, where F-measure is defined as $2 * p * r / (p + r)$.

Table 2. Extract of outdoor images TBox

Countryside_buildings ⊑ ∃contains.Buildings ⊓ ∃contains.Foliage
Foliage ⊔ Grass ⊔ Tree ⊑ Foliage
Rockyside ⊑ ∃contains.Cliff
Roadside ⊑ ∃contains.Road
Coastal ≡ ∃contains.Sea
Forest ⊑ Landscape
Forest ⊑ ∃contains.Foliage
Beach ≡ Coastal ⊓ ∃contains.Sand
Beach ⊑ Natural
Cityscape ⊑ ManMade
∃contains.Sky ⊑ Outdoor
∃contains.Mountain ⊑ Mountainous
Forest ⊓ (Roadside ⊔ Countryside_buildings) ⊑ ⊥
Roadside ⊓ Countryside_buildings ⊑ ⊥
Landscape ⊓ (Mountainous ⊔ Coastal) ⊑ ⊥
Natural ⊓ ManMade ⊑ ⊥

Table 3. Evaluation of analysis and reasoning performance for scene level concepts - Experiment I

Concept	Analysis			Reasoning		
	Recall	Precision	F-M	Recall	Precision	F-M
Indoor	-	-	-	1.00	0.75	0.85
Outdoor	0.99	0.99	0.99	0.99	0.99	0.99
Natural	0.97	0.96	0.97	0.98	0.96	0.97
ManMade	0.18	0.40	0.25	0.18	0.40	0.25
Cityscape	0.18	0.40	0.25	0.18	0.40	0.25
Landscape	0.75	0.63	0.68	0.76	0.68	0.71
Mountainous	0.64	0.28	0.39	0.48	0.30	0.37
Coastal	-	-	-	0.86	0.49	0.63
Beach	0.89	0.26	0.40	0.90	0.31	0.47

Experiment I. In the first experiment, we employed two classifiers at scene level and two classifiers at segment level. The scene classifiers use colour and texture features following a support vector machine [21], and a randomised clustering trees approach [22] respectively. The segment level classifiers use respectively a distance based feature matching approach based on prototypical values [10], and a clustering trees approach [23]. The sets of analysis supported scene and object level concepts are respectively Outdoor, Indoor, Natural, ManMade, Landscape, Beach, Mountainous, Beach and Building, Grass, Foliage, Cliff, Tree, Sea, Sand, Conifers, Boat, Road, Ground, Sky, Trunk, Person. The TBox includes also the concepts Coastal and Cityscape, which are to be entailed through reasoning.

In Tables 3 and 4, the evaluation metrics are given for the scene and segment level concepts respectively. For the case of scene level concepts, we treated the outcome of the analysis based on the semantics of the concepts that the classifiers

Table 4. Evaluation of analysis and reasoning performance for object level concepts - Experiment I

Concept	Analysis			Reasoning		
	Recall	Precision	F-M	Recall	Precision	F-M
Building	0.35	**0.17**	0.22	0.09	**0.83**	0.17
Grass	0.06	**0.40**	0.10	0.01	**0.94**	0.05
Foliage	0.99	0.70	**0.82**	0.90	0.80	**0.85**
Cliff	0.98	0.21	**0.35**	0.54	0.42	**0.47**
Tree	0.22	0.65	0.33	0.18	0.58	0.27
Sand	0.49	0.37	**0.42**	0.92	0.41	**0.56**
Sea	0.72	0.46	**0.56**	0.88	0.49	**0.63**
Conifers	1.00	0.01	0.02	0.50	0.02	0.03
Mountain	0.14	0.01	**0.01**	0.43	0.04	**0.06**
Boat	0.10	0.40	0.16	0.10	0.50	0.17
Road	0.15	0.50	0.23	0.02	0.25	0.03
Ground	**0.06**	0.57	0.19	**0.11**	0.57	0.19
Sky	0.93	0.87	0.89	0.93	0.87	0.89
Trunk	0.38	0.65	0.48	0.38	0.65	0.48
Person	0.49	0.54	0.52	0.49	0.54	0.52

supported. For example, in case the Landscape concept was detected we assumed that the concepts Natural and Outdoor were also detected, although this was not necessarily the case, i.e. the corresponding detectors had no provided a positive outcome. This accounts partially for the low impact of reasoning in the case of scene level descriptions, since the descriptions that were to be inferred through the subsumption axioms that apply to between the scene level concepts, were made explicit. In the case of the Beach and Coastal concepts, where no classifier for Coastal is included, the result of the application of reasoning becomes apparent. A second reason for the relative low affect of reasoning relates to the semantics of the scene concepts themselves. Observing the corresponding TBox (Table 2), ones notices that the scene level concepts are in their majority either atomic concepts or concepts appearing in the left-hand of subsumption axioms, i.e. concepts that cannot be logically entailed through the existence of others.

This is not the case with respect to the object level concepts, where the impact of the reasoning is higher. As illustrated in Table 4, there are cases for which precision is increased, which correspond to concepts involved in disjointness axioms, cases where both recall and precision are improved, which correspond to complex concepts definitions or concepts that appear on the right hand side of subsumption axioms (e.g. Sea, Mountain), and cases where the performance is invariable, which involve concepts not participating in any axiom (e.g. Person) or concepts participating solely in the left hand side of subsumption axioms (e.g. Trunk). Evaluating collectively the performance of analysis amounts to 0.68, 0.49, and 0.57, for recall, precision, and F-measure. The respective values for reasoning are 0.68, 0.63, and 0.65. If we take into consideration only concepts whose semantics are affected by logical associations, the corresponding values become 0.70, 0.64, and 0.67.

Experiment II. In the second experiment, we considered a method based on the combined use of global and local information for the detection of both scene and object level descriptions. Colour, texture and shape descriptors are used, and learning is implemented using support vector machines. The sets of analysis supported scene and object level concepts are respectively Countryside_Buildings, Roadside, Rockyside, Beach and Building, Roof, Grass, Foliage, Dried_Plant,

Table 5. Evaluation of analysis and reasoning performance for scene level concepts - Experiment II

Concept	Analysis			Reasoning		
	Recall	Precision	F-M	Recall	Precision	F-M
Countryside_buildings	0.30	1.0	**0.46**	0.60	0.86	**0.71**
Rockyside	0.68	0.70	**0.69**	0.68	0.79	**0.74**
Roadside	0.68	0.69	0.69	0.68	0.72	0.70
Forest	0.75	0.63	0.69	0.74	0.68	0.71
Coastal	0.85	0.67	0.75	0.86	0.72	0.78
Outdoor	-	-	-	0.99	1.00	0.99
Natural	-	-	-	0.97	1.00	0.98
Mountainous	-	-	-	0.67	0.80	0.74
Beach	-	-	-	0.45	0.76	0.57

Table 6. Evaluation of analysis and reasoning performance for object level concepts - Experiment II

Concept	Analysis			Reasoning		
	Recall	Precision	F-M	Recall	Precision	F-M
Building	0.54	0.69	**0.60**	0.62	0.86	**0.72**
Roof	0.33	0.54	**0.41**	0.33	0.75	**0.46**
Grass	0.49	0.42	0.45	0.30	0.52	0.38
Foliage	0.48	0.84	**0.61**	0.86	0.86	**0.86**
Dried-Plant	0.07	0.11	0.08	0.07	0.13	0.10
Sky	0.95	0.93	0.94	0.95	0.93	0.94
Cliff	0.65	0.45	**0.53**	0.69	0.70	**0.69**
Tree	0.49	0.52	0.51	0.56	0.47	0.51
Sand	0.02	0.10	**0.03**	0.57	0.45	**0.50**
Sea	0.69	0.60	**0.64**	0.85	0.69	**0.76**
Boat	0.41	0.71	0.52	0.33	0.66	0.44
Road	0.50	0.69	**0.58**	0.69	0.71	**0.70**
Ground	0.26	0.33	0.29	0.26	0.33	0.29
Person	0.75	0.51	0.61	0.75	0.51	0.61
Trunk	0.26	0.28	0.27	0.26	0.28	0.27
Wave	0.25	0.5	0.33	0.25	0.5	0.33

Sky, Cliff, Tree, Sea, Sand, Boat, Road, Ground, Person, Trunk, Wave. The concepts Outdoor, Natural, Coastal and Mountainous are to be supported solely through reasoning.

In the Tables 5 and 6, the evaluation metrics are presented for the case of scene and object level concepts respectively. A first observation is the improvement in terms of the descriptions completeness, i.e. the subsumption axioms between the scene level concepts allow to enhance the descriptions supported by analysis and acquire descriptions of more generic concepts such as Natural, and Mountainous. Secondly, we observe that the application of reasoning improves in general the precision of the extracted descriptions. This is a direct outcome of the fact that there exists strong semantic association between the scene and segment level concepts semantics, i.e. there is a significant number of axioms (subsumption and disjointness ones) between them. This explains also the stronger, compared to the first experiment, affect of reasoning in the case of object level descriptions also (Table 6). The aggregated evaluation of scene and object level concepts amounts to 0.37, 0.65, and 0.47 for recall, precision, and F-measure. The respective values for reasoning are 0.77, 0.81, and 0.79. Taking into consideration only concepts which participate to axioms, the corresponding values become 0.29, 0.61, and 0.39 for the case of analysis, and 0.79, 0.82, and 0.81 for the case of reasoning, constituting hence a significant improvement.

6   Relevant Work

In the majority of relevant literature, only crisp DLs approaches have been investigated: in [6], crisp DLs are proposed for inferring descriptions whose semantics lie in logical aggregation, in [8], DLs have been extended with a rule-based approach to realise abductive inference over crisp analysis assertions, while in [11], DLs and rules have been utilised for video annotation using crisp semantics. Fuzzy DLs have been proposed in [24] for the purpose for semantic multimedia retrieval; the fuzzy annotations however are assumed to be available. Fuzzy DLs have been proposed recently in [25] and [26] for enhancing machine learning based extracted image annotations and document classification respectively; however, neither approach addresses the problem of resolving semantic inconsistencies in the initially extracted descriptions.

7   Conclusions

In this paper, we presented a fuzzy DLs based reasoning framework with the aim to enhance the extraction of image semantics through the utilisation of formal semantics. The application of fuzzy DLs semantics allows us to formally address the uncertainty confronted in descriptions extracted through machine learning analysis. Furthermore, through the utilisation of the semantics characterising the available domain knowledge, the proposed reasoning framework addresses and resolves inconsistencies among the initial descriptions. Thereby, and free of assumptions regarding the preceding analysis, it provides the means to integrate descriptions acquired through typical image analysis into a semantically consistent, semantically enhanced annotation. The experiments, though not conclusive, have shown very promising results, indicating that the impact of reasoning is proportional to the level of semantic associations underlying the domain concepts. Future directions include the investigation of extending the framework in order to handle spatial relations semantics, and the combination with probabilistic knowledge as complementary means to handle the uncertainty in semantic image analysis.

Acknowledgements

This work was partially supported by the European Commission under contracts FP6-001765 aceMedia and FP6-507482 KnowledgeWeb.

References

1. Smeulders, A., Worring, M., Santini, S., Gupta, A., Jain, R.: Content-based image retrieval at the end of the early years. IEEE Trans. Pattern Anal. Mach. Intell. 22(12), 1349–1380 (2000)
2. Rao, A., Jain, R.: Knowledge representation and control in computer vision systems. In: IEEE Expert, pp. 64–79 (1988)
3. Draper, B., Hanson, A., Riseman, E.: Knowledge-directed vision: control, learning and integration. Proc. of the IEEE 84(11), 1625–1681 (1996)
4. Little, S., Hunter, J.: Rules-by-example - a novel approach to semantic indexing and querying of images. In: International Semantic Web Conference (ISWC), Hiroshima, Japan, November 7-11, pp. 534–548 (2004)
5. Schober, J.P., Hermes, T., Herzog, O.: Content-based image retrieval by ontology-based object recognition. In: Biundo, S., Frühwirth, T., Palm, G. (eds.) KI 2004. LNCS (LNAI), vol. 3238, Springer, Heidelberg (2004)
6. Neumann, B., Moller, R.: On scene interpretation with description logics (FBI-B-257/04) (2004)
7. Dasiopoulou, S., Mezaris, V., Kompatsiaris, I., Papastathis, V.K., Strintzis, M.G.: Knowledge-assisted semantic video object detection. IEEE Trans. Circuits Syst. Video Techn. 15(10), 1210–1224 (2005)
8. Espinosa, S., Kaya, A., Melzer, S., Möller, R., Wessel, M.: Multimedia interpretation as abduction. In: Proc. International Workshop on Description Logics (DL), Brixen-Bressanone, Italy, June 8-10 (2007)
9. Hollink, L., Little, S., Hunter, J.: Evaluating the application of semantic inferencing rules to image annotation. In: Proc. International Conference on Knowledge Capture (K-CAP), Banff, Alberta, Canada, October 2-5, pp. 91–98 (2005)
10. Petridis, K., Bloehdorn, S., Saathoff, C., Simou, N., Dasiopoulou, S., Tzouvaras, V., Handschuh, S., Avrithis, Y., Kompatsiaris, I., Staab, S.: Knowledge representation and semantic annotation of multimedia content. In: IEE Proceedings on Vision Image and Signal Processing, Special issue on Knowledge-Based Digital Media Processing 153 (June 2006)
11. Bagdanov, A., Bertini, M., DelBimbo, A., Serra, G., Torniai, C.: Semantic annotation and retrieval of video events using multimedia ontologies. In: Proc. IEEE International Conference on Semantic Computing (ICSC), Irvine, CA, USA (2007)

12. Dasiopoulou, S., Heinecke, J., Saathoff, C., Strintzis, M.: Multimedia reasoning with natural language support. In: Proc. IEEE International Conference on Semantic Computing (ICSC), Irvine, CA, USA, September 17-19 (2007)
13. Straccia, U.: A fuzzy description logic for the semantic web. In: Sanchez, E. (ed.) Fuzzy Logic and the Semantic Web. Capturing Intelligence, pp. 73–90. Elsevier, Amsterdam (2006)
14. Stoilos, G., Stamou, G., Tzouvaras, V., Pan, J., Horrocks, I.: The fuzzy description logic f-SHIN. In: International Workshop on Uncertainty Reasoning For the Semantic Web (URSW), Galway, Ireland, November 7 (2005)
15. Baader, F., Calvanese, D., McGuinness, D.L., Nardi, D., Patel-Schneider, P.F. (eds.): The description logic handbook: Theory, implementation, and applications. In: Description Logic Handbook. Cambridge University Press, Cambridge (2003)
16. Stoilos, G., Stamou, G., Pan, J.: Handling imprecise knowledge with fuzzy description logic. In: Proc. International Workshop on Description Logics (DL), Lake District, UK (2006)
17. Straccia, U.: Reasoning within fuzzy description logics. J. Artif. Intell. Res (JAIR) 14, 137–166 (2001)
18. Straccia, U.: Transforming fuzzy description logics into classical description logics. In: Proc. European Conference on Logics in Artificial Intelligence (JELIA), Lisbon, Portugal, September 27-30, pp. 385–399 (2004)
19. Haase, P., van Harmelen, F., Huang, Z., Stuckenschmidt, H., Sure, Y.: A framework for handling inconsistency in changing ontologies. In: Proc. of International Semantic Web Conference (ISWC), Galway, Ireland, November 6-10, pp. 353–367 (2005)
20. Kalyanpur, A., Parsia, B., Sirin, E., Grau, B.C.: Repairing unsatisfiable concepts in owl ontologies. In: Proc. of European Semantic Web Conference (ESWC), Budva, Montenegro, June 11-14, pp. 170–184 (2006)
21. LeBorgne, H., Guérin-Dugué, A., O'Connor, N.E.: Learning midlevel image features for natural scene and texture classification. IEEE Trans. Circuits Syst. Video Techn. 17(3), 286–297 (2007)
22. Moosmann, F., Triggs, B., Jurie, F.: Randomized clustering forests for building fast and discriminative visual vocabularies. In: Neural Information Processing Systems (NIPS) (November 2006)
23. Moller, R., Neumann, B., Wessel, M.: Towards computer vision with description logics: Some recent progress. In: Proceedings Integration of Speech and Image Understanding, Corfu, Greece, pp. 101–115 (1999)
24. Umberto, S., Giulio, V.: Dlmedia: an ontology mediated multimedia information retrieval system. In: Proc. International Workshop on Description Logics (DL), Brixen-Bressanone, Italy, June 8-10 (2007)
25. Simou, N., Athanasiadis, T., Tzouvaras, V., Kollias, S.: Multimedia reasoning with f-shin. In: 2nd International Workshop on Semantic Media Adaptation and Personalization, London, UK (2007)
26. Mylonas, P., Simou, N., Tzouvaras, V., Avrithis, Y.: Towards semantic multimedia indexing by classification and reasoning on textual metadata. In: Knowledge Acquisition from Multimedia Content Workshop, Genova, Italy (2007)

Enriching a Thesaurus to Improve Retrieval of Audiovisual Documents

Laura Hollink, Véronique Malaisé, and Guus Schreiber

Free University Amsterdam
de Boelelaan 1081a
1081 AH Amsterdam, The Netherlands

Abstract. In many archives of audiovisual documents, retrieval is done using metadata from a structured vocabulary or thesaurus. In practice, many of these thesauri have limited or no structure. The objective of this paper is to find out whether retrieval of audiovisual resources from a collection indexed with an in-house thesaurus can be improved by anchoring the thesaurus to an external, semantically richer thesaurus. We propose a method to enrich the structure of a thesaurus and we investigate its added value for retrieval purposes.

We first anchor the thesaurus to an external resource, WordNet. From this anchoring we infer relations between pairs of terms in the thesaurus that were previously unrelated. We employ the enriched thesaurus in a retrieval experiment on a TRECVid 2007 dataset. The results are promising: with simple techniques we are able to enrich a thesaurus in such a way that it adds to retrieval performance.

1 Introduction

The objective of this paper is to investigate whether retrieval of audiovisual documents that are indexed with an in-house thesaurus can be improved by anchoring the thesaurus to an external, semantically richer thesaurus.

Many collections of audiovisual documents are indexed manually with the help of a local thesaurus. The manual indexing process is time-consuming, therefore the tendency is to only use a small set of terms to describe a document. The annotations are usually of high quality. We point out that the opposite can be said about automatic annotation using content-based feature detectors. This approach results in many annotations, but their quality is unreliable.

A low number of annotations per document can lead to low recall of search results. One way to overcome this issue is query expansion, where documents are retrieved not only with the initial query term, but also with closely related terms [18]. In the context of concept-based search, where queries are posed in terms of thesaurus concepts, query expansion depends on a rich thesaurus structure. However, local thesauri are often limited in breadth and depth. In this paper we report on an experiment in which we enrich a local thesaurus and study its added value for retrieval.

D. Duke et al. (Eds.): SAMT 2008, LNCS 5392, pp. 47–60, 2008.

The study is performed on a dataset of the Netherlands Institute for Sound and Vision (Sound & Vision). The institute stores over 700,000 hours of Dutch broadcast video, and archives every day the daily broadcast in digital format. It has an in-house thesaurus, the GTAA, with limited structure, which is used to index and search the collection.

Our approach consists of two steps. First, we anchor the GTAA thesaurus to an external resource, the English-American WordNet [4], by searching for related concepts (*synsets* in WordNet) using a syntactic alignment procedure. The alignment is based on lexical comparison of term descriptions in the two resources. Such a mainly lexical alignment approach is bound to be incomplete and at times incorrect. However, considering the state of the art of ontology alignment, this is a realistic situation [3]. Second, based on this anchoring, we enrich the in-house thesaurus by inferring potential new relations between terms within the thesaurus.

To investigate the value of the enriched thesaurus for retrieval purposes, we perform an experiment in which we compare retrieval results achieved with the in-house thesaurus to results obtained with the enriched thesaurus. The experiment is performed on a part of the collection of Sound & Vision that was used in the TRECVid 2007 conference [15]. We use the queries and ground truth provided by TRECVid. In addition, Sound & Vision kindly provided us with the metadata of this dataset in the form of manual annotations of the audiovisual documents with GTAA terms.

Our hypothesis is that anchoring the in-house thesaurus to a rich external source will help retrieval, particularly with respect to recall: the richer semantic structure should lead to more matches. We are interested in finding out how much this approach jeopardizes precision and whether the joint effect can be judged to be positive or negative.

This paper is structured as follows. Section 2 describes the GTAA and its anchoring to WordNet. In Section 3 we describe how new thesaurus relations are inferred from the GTAA-WordNet links. Section 4 describes the setup of the retrieval experiment and the TRECVid dataset. The results of the retrieval experiment are analyzed in Section 5. We conclude with a discussion and directions for future work in Section 6.

2   Anchoring the GTAA Thesaurus to WordNet

2.1   The GTAA Thesaurus

The GTAA is a Dutch, faceted thesaurus resulting from the merging of several controlled vocabularies used by audiovisual archives in the Netherlands. Its name is a Dutch acronym for "Common Thesaurus for Audiovisual Archives". At the Netherlands Institute for Sound and Vision it is used for manual annotation of the extensive collection of broadcast video.

The GTAA thesaurus contains approximately 160,000 terms, organised in six facets: Location, Person name, Name, Maker, Genre and Subject. Location describes either the place(s) where the video was shot, the places mentioned or seen on the screen or the places the video is about. **Person name** is used for people who are either seen or heard in the video, or who are the subject of the program; **Name** has the same function for named organisations, groups, bands, periods, events, etc. **Maker** and **Genre** describe the creators and genre of a TV program. The **Subject** facet is used to describe what a program is about, and aims to contain terms for all topics that could appear on TV, which makes its scope quite broad.

Since our aim is to retrieve video based on what it is about, the focus of the present paper is on the Subject facet. In the future we intend to include also other facets, such as Locations and Names. The Subject facet contains 3878 terms[1]. It is organised according to the semantic relations defined in the ISO-standard 2788 for thesauri [8], namely **Broader Term** (linking a specialized concept to a more general one), its inverse relation **Narrower Term** (linking a general concept to a more specialized one), and **Related Term** (denoting an associative link). The GTAA contains 3,382 broader/narrower relations and 7,323 associative relations between terms in the subject facet. The broader/narrower hierarchy is shallow: 85% of the hierarchy is no more than three levels deep. For integration purposes, we used a version of the GTAA that was converted to SKOS in an earlier effort [1]. SKOS provides a common data model to represent thesauri using RDF and port them to the Semantic Web [12].

2.2 Anchoring to WordNet

The archives of the Netherlands Institute for Sound and Vision are searched by broadcast professionals, who reuse material to create new television programs, and by the general public. Querying large audiovisual archives remains difficult. A recognised approach to increase recall is query expansion: retrieving documents with not only the initial query term, but also closely related terms [18]. If the aim is to use thesaurus relations for the expansion (as opposed to using lexical techniques), a rich thesaurus with many interrelated terms is necessary. To increase the structure of the GTAA, we anchor it to an external thesaurus, WordNet, and take advantage of its rich semantic structure.

WordNet is a lexical database of the English language. It currently contains 155,287 English words: nouns, verbs, adjectives and adverbs. Many of these words are polysemous, which means that one word has multiple meanings or senses. The word 'tree', for example, has three word-senses: tree#1 (woody plant), tree#2 (figure) and Tree#3 (English actor). WordNet distinguishes 206,941 word-senses.

Word-senses are grouped into synonym sets (synsets) based on their meaning and use in natural language. Each synset represents one distinct concept. An example of a synset is cliff#1, drop#4, drop-off#2, described as "a steep high face of rock". Semantic relations and lexical relations exist between word-senses and between synsets. For the purpose of this paper we will not go into details of all these relations, but rather explain the most common ones. The main hierarchy in WordNet is built on hypernym/hyponym relations between synsets,

---

[1] In addition to the 3,878 preferred terms, the subject facet contains around 2,000 non-preferred terms.

which are similar to superclass/subclass relations. Other frequent relations are meronym and holonym relations, which denote part-of and whole-of relations respectively. Each synset is accompanied by a 'gloss': a definition and/or some example sentences.

WordNet is freely available from the Princeton website[2]. In addition, W3C has released a RDF/OWL representation of WordNet version 2.0[3]. In this study we use this RDF/OWL version, as it allows us to use Semantic Web tools and standards to query the WordNet database.

Anchoring GTAA to WordNet is non-trivial, since the two thesauri are in different languages. As a first step, we used a Dutch lexical database (Celex) to find alternative forms of the terms in the thesaurus. In addition to the original preferred terms and non-preferred terms, we added singular forms (since WordNet words are mostly singular) and synonyms. Compound terms were split into separate words (again using Celex), which were also added. All forms, the original ones as well as the newly added ones, were used for the anchoring to WordNet as this increases the possible coverage of the anchoring.

Second, we queried an online bilingual dictionary[4] for the Dutch terms, which provided English translations and one-sentence descriptions. Third, we anchored the English GTAA terms to WordNet. In contrast to many anchoring methods (e.g. [9]), we do not compare the terms from the two thesauri, but measure the lexical overlap of their descriptions. The same approach has been followed by Knight [10]. This approach is especially well suited in our case since, much to our surprise, the one-sentence descriptions of the online dictionary *are* the WordNet glosses for 99% of the words. The anchoring process is described in more detail in [11].

GTAA terms that were found to correspond to multiple WordNet synsets were anchored to all those synsets. There were three reasons why we didn't attempt to do sense disambiguation. First, we are aiming for an increased recall so our primary focus is finding correct correspondences rather than avoiding incorrect correspondences. Second, disambiguation of terms with little context (which is the case for the GTAA terms) is difficult. In the future, we intend to take into account broader terms for disambiguation purposes. The third and most important reason is that linking to more than one synset is often correct because WordNet makes finer distinctions than the GTAA. For example, WordNet distinguishes four meanings for the GTAA term "chicken", described by the glosses: 'adult female chicken', 'the flesh of a chicken used for food', 'a domestic fowl bred for flesh or eggs' and 'a domesticated gallinaceous bird thought to be descended from the red jungle fowl'. This fine-grained distinction is absent in the GTAA.

In total, 1,855 GTAA terms were anchored to WordNet. 885 of those corresponded to only one synset, 464 corresponded to two synsets, 242 to three synsets, 121 to four, 76 to five, 35 to six and 32 corresponded to seven or more synsets. Some WordNet synsets are linked to more than one GTAA term. For

---

[2] http://wordnet.princeton.edu/
[3] http://www.w3.org/TR/wordnet-rdf/
[4] http://lookwayup.com

example, the WordNet synset "Studio" was an anchor for both the GTAA terms "Atelier" and "Studio". An informal evaluation of a small sample of the correspondences suggests that the number of synsets that is aligned with a particular GTAA term is not an indication of the quality of the matches; GTAA terms that are matched to multiple synsets are equally well matched as GTAA terms that are matched to only one synset.

Correspondences based on split compound words are less good than those based on original preferred terms or singular forms. However, we estimate that only 10% is actually incorrect. The majority anchors a term to a related or broader synset.

3 Thesaurus Enrichment

3.1 Approach

We used the anchoring to WordNet to infer new relations within the GTAA. Using SeRQL [2] queries we related pairs of GTAA subject terms that were not previously related. Figure 1 illustrates how a relation between two terms in the GTAA, t_1 and t_2, is inferred from their correspondence to WordNet synsets w_1 and w_2. If t_1 corresponds to w_1 and t_2 corresponds to w_2, and w_1 and w_2 are closely related, we infer a relation between t_1 and t_2. The inferred relation is symmetric, illustrated by the two-way arrow between t_1 and t_2.

Two WordNet synsets w_1 and w_2 are considered to be 'closely related' if they are connected though either a direct (i.e. one-step) relation without any intermediate synsets or an indirect (i.e. two-step or three step) relation with one or two intermediate synsets. The latter situation is shown in Figure 1. From all WordNet relations, we used only meronym and hyponym relations, which roughly translate to part-of and subclass relations, and their inverses holonym and hypernym. A previous study demonstrated that other types of WordNet relations do not improve retrieval results when used for query expansion [7]. Both meronym and hyponym can be considered hierarchical relations in a thesaurus. Only sequences of two relations are included in which each has the same direction, since previous research showed that changing direction, especially in the

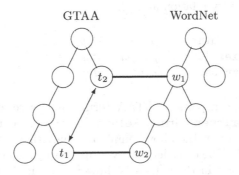

Fig. 1. Using the anchoring to WordNet to infer relations within the GTAA

hyponym/hypernym hierarchy, decreases semantic similarity significantly [7,5]. For example, w_a hypernym of w_b hyponym of w_c is not included.

3.2   Newly Inferred Relations

A total of 1039 pairs of GTAA terms was newly related: 404 with one step between WordNet synsets w_1 and w_2, 362 with 2 steps and 273 with three steps between w_1 and w_2. Around 90% of the relations were derived from hyponym relations and only 10% from meronym relations, which is a more rare relation in WordNet.

Although we intend to only implicitly evaluate the quality of the inferred relations by looking into their value for retrieval, a manual inspection of a portion of the new relations suggests that many of them have the potential to be beneficial for retrieval. At the same time, many others seem so trivial that we don't expect much added value. Only very few seem wrong. We did not detect a difference in quality between relations inferred from hyponyms and those inferred from meronyms. Examples of new relations that we consider valuable are:

```
squid            - colouring (pigment)  (hyponym 1 step)
pharmacy         - medicine             (hyponym 1 step)
barbecues        - picknicks            (hyponym 2 steps)
national anthems - music                (hyponym 3 steps)
pearls           - jewelery             (hyponym 3 steps)
fjords           - seas                 (meronym 1 step)
cement           - concrete             (meronym 2 steps)
flour            - meal                 (meronym 3 steps)
```

Relations that we consider trivial are, for example:

```
cigarettes  - cigars       (meronym 1 step)
computers   - machines     (hyponym 1 step)
coffeeshops - restaurants   (hyponym 1 step)
```

Examples of incorrect new relation are:

```
acupuncture - negotiation  (hyponym 1 step)
banknotes   - copies       (hyponym 1 step)
apples      - foetusses     (hyponym 2 step)
```

Inferred relations between pairs of GTAA terms that were already each others Broader Term, Narrower Term or Related Term were not included in the retrieval experiment, nor in the above numbers, since they do not *add* to the structure of the GTAA. The considerable overlap between what we inferred and what was already present in the GTAA is, however, an indication that the inferred relations make sense.

4   Retrieval with the Enriched Thesaurus

We employed the enriched thesaurus for retrieval of television programs from the archives of the Netherlands Institute for Sound and Vision. The programs were annotated with subject terms from the GTAA. Our aim is twofold. First, we want to know the value of the inferred relations for retrieval, and compare that to retrieval with existing GTAA relations. Second, we are interested in the added value of the inferred relations when we use them in combination with the existing GTAA relations.

4.1   Experimental Setup

We query the collection in nine runs, each using a different type of relation or combination of relations:

Exact. Only programs annotated with the query term are returned. This run is used as a baseline.

GTAA bro. Programs annotated with the query term or broader terms are returned.

GTAA nar. Programs annotated with the query term or narrower terms are returned.

GTAA rel. Programs annotated with the query term or related terms are returned.

GTAA all. Programs annotated with the query term or terms that are related through (a combination of) GTAA relations (narrower, broader, related) are returned.

Via WN 1 step. Programs annotated with the query term or terms related through a one-step inferred relation are returned.

Via WN 2 step. Programs annotated with the query term or terms that are related through a two-step inferred relation are returned.

Via WN 3 step. Programs annotated with the query term or terms that are related through a three-step inferred relation are returned.

Via WN all. Programs annotated with the query term or with a term that is related through a (combination of) one-, two- or three-step inferred relations are returned.

All. Programs annotated with the query term or terms that are related in any of the above ways are returned.

At present, we allowed three steps between the query term and the target term. More than three steps resulted in an explosion of the number of returned documents.

Of each run, we measure the precision (Prec), recall (Rec) and the harmonic mean of the two, called F_1-measure:

$$\text{Prec} = \frac{|\text{Retrieved\&Relevant}|}{|\text{Retrieved}|} \qquad \text{Rec} = \frac{|\text{Retrieved\&Relevant}|}{|\text{Relevant}|}$$

$$F_1 = 2 \cdot \frac{\text{Prec} \cdot \text{Rec}}{\text{Prec} + \text{Rec}}$$

where |Retrieved| is the number of programs a run retrieved, and |Relevant| is the number of programs that is relevant for a query.

4.2   TRECVid Data: Corpus and Queries

In order to determine the added value of the inferred relations, a dataset and a ground truth are needed that are large enough to distinguish if there are any significant differences between runs. In the current study, we used the TRECVid 2007 dataset for the high-level feature extraction task. This dataset consists of 50 hours of news magazine, science news, news reports, documentaries, educational programming, and archival video from the Netherlands Institute for Sound and Vision, 36 queries ('features') and a manually constructed ground truth. Sound & Vision kindly provided us with the metadata of this dataset in the form of manual annotations of the television programs with GTAA terms.

The queries consist of a single or moderately complex query term, such as **Sports** or **Explosion_Fire**. This corresponds to the types of queries that are posed in the online search interface of Sound & Vision, where the majority of queries consist of a single term, sometimes completed with a broadcast date. Simple, unequivocal queries are a requirement in this type of study, as complex queries could obscure the results.

We manually translated the high-level features to get queries in terms of GTAA subjects. Features that consisted of two subjects were interpreted as the union of both and we queried for programs containing one and/or the other. This was clearly the intended semantics of the features as can be seen from descriptions such as the one for **Walking_Running**: Shots depicting a person walking or running. Of the initial 36 features, three did not have a satisfactory translation, and were therefore discarded.

All TRECVid tasks are at the level of shots, while the GTAA subject annotations are at the level of television programs. We adapted the given ground truth to be on program-level. In the resulting ground truth, nine queries appeared in more than 2/3 of the programs and were therefore discarded. **Person** and **Face**, for example, appeared in each program. Six programs were not usable in the present experiment since they did not have a subject annotation and could therefore never be retrieved.

After adaptation, the dataset consisted of 104 television programs annotated with on average 3.6 GTAA subject terms, 25 queries and a ground truth that listed on average 27 relevant programs for each query.

We stress that although we use the TRECVid dataset, our results are not comparable to those of systems that participated in the TRECVid 2007 conference. We retrieve programs based on metadata and the structure of a thesaurus, while TRECVid participants retrieve shots based on the audiovisual signal.

5   Results and Interpretation

Table 1 and Figure 2 summarize the results. Please note that although the range of the y-axis of the plots in Figure 2 differ, the height of the bars represents the same value in all three plots.

Throughout this section we use Students paired t-test to compare the performance of runs[5]. Significance levels, t-values, degrees of freedom and the appropriate version of the test (one or two tailed) will be reported as, for example, $(t =, p =, df =$, one-tailed).

5.1   Existing GTAA Relations

The results of the runs using existing thesaurus relations merely confirm what was known about thesaurus based retrieval. We discuss them since they form a baseline against which we can compare the performance of the inferred relations. The human entered subject terms are reliable, and using them gives high precision, in our case even 100% (the 'exact' run). We suspect that the level of correctness of our annotations was higher than usual thanks to the special attention the Netherlands Institute for Sound and Vision gave to the collection they prepared for TRECVid. In many cases, of course, human annotators do err and disagree [17]. The time-consuming nature of human annotation causes the number of subject terms per program to be low, much lower than the number of topics that is visible in the video. This makes the recall of the run that relies solely on these human annotations unacceptably low: 2% on average.

Table 1. Precision, recall and F_1-measure of the nine runs, summarized by the mean $\pm$ the standard deviation

Run	Precision	Recall	F_1
GTAA exact	1.00 ± 0.00	0.03 ± 0.06	0.17 ± 0.11
GTAA broader	0.81 ± 0.35	0.03 ± 0.06	0.16 ± 0.10
GTAA narrower	0.89 ± 0.30	0.04 ± 0.07	0.20 ± 0.12
GTAA related	0.42 ± 0.26	0.33 ± 0.23	0.29 ± 0.15
GTAA all	0.39 ± 0.25	0.46 ± 0.27	0.34 ± 0.17
Via WN one-step	0.70 ± 0.40	0.05 ± 0.07	0.18 ± 0.10
Via WN two-step	0.76 ± 0.30	0.07 ± 0.11	0.20 ± 0.12
Via WN three-step	0.81 ± 0.38	0.04 ± 0.06	0.16 ± 0.10
Via WN all	0.58 ± 0.40	0.13 ± 0.17	0.26 ± 0.14
All	0.38 ± 0.25	0.57 ± 0.29	0.38 ± 0.19

---

[5] The t-test requires a normal distribution. Normality was assessed with Quantile-Quantile plots. Although for some of the smaller samples - the exact run, for example, returned programs for only seven queries and had therefore only 7 precision values - normality could not be proven, we assume that precision and recall are normally distributed quantities given a large number of queries.

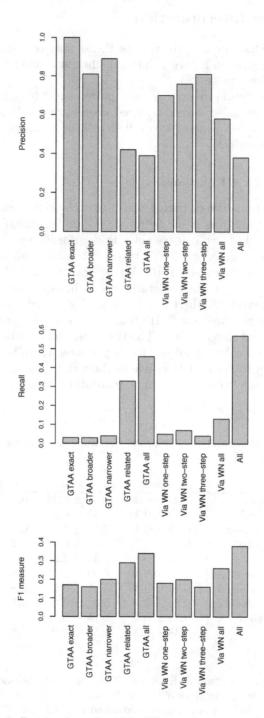

Fig. 2. Retrieval results with different thesaurus relations

Including terms that are broader than the query does not add to recall. This is partly due to the fact that our queries are all fairly general, and many don't have a broader term. Still, it is a confirmation of what was found in an earlier study [7]. Narrower terms, on the other hand, do seem to add a little to recall, although the result is not statistically significant ($t =1.51$, $p =0.07$, $df =24$, one-sided), and they maintain a high precision. This is what we would expect from the definition of narrower terms: "the scope (meaning) of one falls completely within the scope of the other" [13]. Related terms are less reliable: precision halves compared to using only exact matches ($t =7.14$, $p <0.01$, $df =7$, two-sided), but recall increases to 33% ($t =6.63$, $p <0.01$, $df =24$, one-sided).

Combining the hierarchical broader/narrower relations with the related terms, only slightly (but significantly) lowers precision further compared to using only the related terms ($t =1.9$, $p =0.03$, $df =24$, two-sided). It does, however, raise recall to 46% ($t =4.3$, $p <0.01$, $df =24$, one-sided). This suggests that also sequences of different types of relations are beneficial to retrieval.

5.2   Newly Inferred Relations

The one-, two- and three-step inferred relations perform equally well. We observe a small difference in precision (three-step is higher), but this is not significant ($t =1.3$, $p =0.24$, df = 7, two-sided). This suggests that the notion of relatedness can be interpreted in a broad sense and does not need to be restricted to only one step in the WordNet hierarchy.

When the one-step, two-step and three step inferred relations are combined ('Via WordNet All') precision remains relatively high: 58%. Recall, on the other hand, is low: 13%. On the whole, the inferred relations give results that are comparable to the results of existing relations in the GTAA that were created by experts. When comparing them to GTAA narrower terms, they score better on recall but worse on precision. When comparing them to GTAA related terms we observe the opposite effect: the inferred relations score higher on precision but lower on recall. This difference in recall can in part be explained from the fact that there are 7 times as many related terms as inferred relations. Also with respect to F_1-measures, the performance of the inferred relations is between that of GTAA narrower and GTAA related terms. There were no significant differences between the F_1 of inferred relation and GTAA narrower terms ($t =0.9$, $p =0.40$, df = 8, two-sided) or GTAA related terms ($t =2.0$, $p =0.07$, df = 12, two-sided). These results show that the inferred relations are valuable for retrieval in situations where there is no other structure in the vocabulary.

Using all relations together improves the recall significantly over using only the existing GTAA relations ($t =4.0$, $p <0.01$, $df =24$, one-sided). Again, this suggests that combination of different types of relations is beneficial to the retrieval results. It also suggests that enrichment of a weakly structured thesaurus has added value to the retrieval results.

The mean increase in recall from the 'GTAA all' run to the 'All' run was 0.11. This increase could in part be attributed to the higher number of retrieved programs. However, the increase in recall was significantly more than we would expect if the additionally retrieved programs were randomly taken from the collection (t =2.40, p =0.02, df =27, two-sided). We calculated the expected increase in recall $\mathbb{E}_{incr}$ for each query as follows:

$$\mathbb{E}_{incr} = \frac{(R_{All} - R_{GTAAall}) \cdot (C - RC_{GTAAall})}{N - R_{GTAAall}} \cdot \frac{1}{C}$$

where R_{All} and $R_{GTAAall}$ are the number of retrieved programs in the 'All' and 'GTAA all' runs respectively, C is the number of correct programs for the query in the collection, $RC_{GTAAall}$ is the number of correctly retrieved programs in the 'GTAA all' run and N is the total number of programs in the collection (104 in our case).

6   Discussion and Future Work

In this paper we experimented with retrieval using a thesaurus that was enriched by anchoring it to an external resource. We have shown that with simple techniques new relations can be inferred that are valuable for retrieval purposes.

We investigated both the effect of only using the newly inferred relations and using them in combination with existing thesaurus relations. Retrieval with only the inferred relations yielded an F_1-measure of around 0.2. This is comparable to the intuitive and widely used approach of using `Narrower Term` thesaurus relations. This finding suggests that it is possible to use an external resource to enrich an otherwise unstructured vocabulary and base the retrieval on the inferred relations. For example, we see possibilities to enrich lexicons of high-level feature detectors that are used in content-based video retrieval. LSCOM is such a vocabulary for annotation and retrieval of video, containing concepts that represent realistic video retrieval problems, are observable and are (or will be) detectable with content-based video retrieval techniques [14]. In a recent effort, LSCOM was manually linked to the CyC knowledge base[6], thus creating structure within LSCOM. We would be interested to compare and combine this manually added structure with an enriched version of LSCOM using the methods proposed in the present paper.

When the inferred relations are used in combination with existing thesaurus relations, it appears that the use of the enriched set of relations increases recall moderately (from 0.46 in the 'GTAA all' run to to 0.57 in the 'all' run) with comparable precision (0.39 vs. 0.38). This indicates that it is beneficial to enrich an already structured thesaurus. However, the number of additionally retrieved documents was too low to draw any definite conclusions about the added value of the inferred relations over an already structured thesaurus. We suspect that a higher number of inferred relations will be needed to confirm this finding. Future research directions therefore include alternative methods of thesaurus enrichment.

When looking at the F-measure scores it is interesting to note that mixing relationships increases performance, both in the non-enriched ("GTAA all") and

---

[6] http://www.cyc.com/

in the enriched ("all") case. This suggests that the nature of the relationship (broader, narrower, related) is not a big issue, at least not in this case. It would be worthwhile to study this in more detail in future experiments.

In future work we want to consider the effect of the inferred relations in more detail. Further experiments could reveal what the effect of different WordNet relations is: hyponyms, meronyms, but also other relations that were not yet used. A further distinction between types of inferred relations, which could start at the anchoring phase, will give more insight into the optimal ranking strategies for semantic search results.

The use of TRECVid data enabled us to experiment on a dataset of reasonable size. However, it also raises some issues. The TRECVid ground truth is based on the pooled results of TRECVid 2007 participants: only items returned by at least one of the participants are judged, while items not returned by anyone are considered incorrect. This could in theory lead to a negative image of our results, since we did not contribute to the pool. It is been argued that this is a negligible problem. Zobel [19], for example, demonstrated that the difference in rating between a system in- and outside the pool is small. However, all systems in his test were content-based image retrieval systems. The retrieval approach under consideration in the present paper is concept-based. Since we use another type of information (metadata instead of the audiovisual signal), it is well possible that we retrieve a set of documents that is disjoint from the set that was retrieved by the content-based systems that contributed to the pool. Therefore, the effect of being outside the pool is potentially larger.

The translation of TRECVid topics to GTAA terms was done manually. In a final application this translation would be done either automatically, which is done in [16], or by the searcher, as in [6]. However, in the present paper our goal was not to build an application but to investigate the possibilities of retrieval with an automatically enriched thesaurus.

References

1. van Assem, M., Malaisé, V., Miles, A., Schreiber, A.T.: A method to convert thesauri to skos. In: Proceedings of the Third European Semantic Web Conference, Budvar, Montenegro, pp. 95–109 (2006)
2. Broekstra, J., Kampman, A.: SeRQL: A second generation RDF query language. In: Proceedings of the SWAD-Europe Workshop on Semantic Web Storage and Retrieval, Amsterdam, The Netherlands, pp. 13–14 (November 2003)
3. Euzenat, J., Isaac, A., Meilicke, C., Shvaiko, P., Stuckenschmidt, H., Šváb, O., Svátek, V., van Hage, W.R., Yatskevich, M.: First results of the ontology alignment evaluation initiative 2007. In: Ashpole, B., Ehrig, M., Euzenat, J., Stuckenschmidt, H. (eds.) Ontology Matching, CEUR Workshop Proc. (2007)
4. Fellbaum, C. (ed.): WordNet: an electronic lexical database. MIT Press, Cambridge (1998)
5. Hirst, G., St-Onge, D.: Lexical chains as representations of context for the detection and correction of malapropisms. In: Fellbaum, C. (ed.) WordNet: An Electronic Lexical Database, pp. 305–332. MIT Press, Cambridge (1998)

6. Hollink, L., Schreiber, A.T., Wielemaker, J., Wielinga, B.J.: Semantic annotation of image collections. In: Proceedings of the K-Cap 2003 Workshop on Knowledge Markup and Semantic Annotation (October 2003)
7. Hollink, L., Schreiber, G., Wielinga, B.: Patterns of semantic relations to improve image content search. Journal of Web Semantics 5, 195–203 (2007)
8. International Organization for Standardization. ISO 2788:1986. Guidelines for the establishment and development of monolingual thesauri. ISO, Geneva (1986)
9. Khan, L.R., Hovy, E.: Improving the precision of lexicon-to-ontology alignment algorithm. In: AMTA/SIG-IL First Workshop on Interlinguas, San Diego, CA, USA (October 1997)
10. Knight, K., Luk, S.: Building a large-scale knowledge base for machine translation. In: The AAAI 1994 Conference (1994)
11. Malaisé, V., Isaac, A., Gazendam, L., Brugmann, H.: Anchoring dutch cultural heritage thesauri to wordnet: two case studies. In: ACL 2007 Workshop on Language Technology for Cultural Heritage Data (2007)
12. Miles, A., Bechhofer, S.: SKOS simple knowledge organization system reference. W3C working draft (January 25, 2008) Electronic document (accessed, April 2008), http://www.w3.org/TR/skos-reference/
13. Miles, A., Brickley, D.: SKOS core guide. W3C working draft (November 2005). Electronic document (accessed, February 2008), http://www.w3.org/TR/swbp-skos-core-guide/
14. Naphade, M., Smith, J.R., Tesic, J., Chang, S.-F., Hsu, W., Kennedy, L., Hauptmann, A., Curtis, J.: Large-scale concept ontology for multimedia. IEEE MultiMedia 13(3), 86–91 (2006)
15. Over, P., Kraaij, W., Smeaton, A.F.: TRECVID 2007 - an introduction. In: TREC Video Retrieval Evaluation Online Proceedings (2007)
16. Snoek, C.G.M., Huurnink, B., Hollink, L., de Rijke, M., Schreiber, G., Worring, M.: Adding semantics to detectors for video retrieval. IEEE Transactions on Multimedia 9(5), 975–986 (2007)
17. Volkmer, T., Thom, J.A., Tahaghoghi, S.M.M.: Exploring human judgement of digital imagery. In: ACSC 2007: Proceedings of the thirtieth Australasian conference on Computer science, Darlinghurst, Australia, pp. 151–160. Australian Computer Society, Inc. (2007)
18. Voorhees, E.: Query expansion using lexical-semantic relations. In: Croft, W.B., van Rijsbergen, C.J. (eds.) Proceedings of the 17th Annual International ACM SIGIR Conference on Research and Development in Information Retrieval, New York, NY, USA, pp. 61–69. Springer, Heidelberg (1994)
19. Zobel, J.: How reliable are the results of large-scale information retrieval experiments? In: SIGIR 1998: Proceedings of the 21st annual international ACM SIGIR conference on Research and development in information retrieval, pp. 307–314. ACM, New York (1998)

Tag Suggestr: Automatic Photo Tag Expansion Using Visual Information for Photo Sharing Websites

Onur Kucuktunc, Sare G. Sevil, A. Burak Tosun, Hilal Zitouni,
Pinar Duygulu, and Fazli Can

Bilkent University, Department of Computer Engineering, Ankara 06800, Turkey

Abstract. In this paper, we propose an automatic photo tag expansion
system for the community photo collections, such as Flickr[1]. Our aim is
to suggest relevant tags for a target photograph uploaded to the system
by a user, by incorporating the visual and textual cues from other related photographs. As the first step, the system requires the user to add
only a few initial tags for each uploaded photo. These initial tags are
used to retrieve related photos including the same tags in their tag lists.
Then the set of candidate tags collected from a large pool of photos is
weighted according to the similarity of the target photo to the retrieved
photo including the tag. Finally, the tags in the highest rankings are
used to automatically expand the tags of the target photo. The experimental results on Flickr photos show that, the use of visual similarity of
semantically relevant photos to recommend tags improves the quality of
suggested tags compared to only text-based systems.

1 Introduction

Recently, large number of photos have become available in photo sharing services.
Although the advances in content based image retrieval studies are promising
[12], scaling these techniques to web is difficult. On the other hand, users are
willing to annotate the images manually [8] allowing the tag based search systems
to be practical.

In the community photo collections, such as Flickr, tags are generally assigned
by users who upload the photos, identifying the location (place, country, etc.)
where the photo is taken, as well as the objects/people appearing in the image, together with some specific words related to camera characteristics, interest
groups etc. However, the tags are usually subjective, noisy and in a limited number, reducing the accessibility of the photographs. Tag suggestion systems, that
can provide related tags to be selected, are therefore important to eliminate the
limitations and to guide the users.

Our motivation in this work is to provide a system that enhances Flickr's
search capabilities by automatically recommending meaningful tags for annotating photographs. We propose a tag suggestion system which expands the tags

---

[1] http://www.flickr.com

D. Duke et al. (Eds.): SAMT 2008, LNCS 5392, pp. 61–73, 2008.

of an image provided by the user, incorporating the tags of the other images which are visually similar. The main contribution of our approach lies in the use of visual information, unlike the previous studies which focuses only on textual information.

First, a few already existing tags (in the order of 2 or 3) are used to form an initial query to find related images including these tags. Then, all the other tags co-occurring with these images are listed as the candidate recommendations. The tags are then weighted according to the similarity of the images, resulting in a higher ranking for the tags coming from visually similar images.

The rest of the paper is organized as follows: related work is discussed in Section 2, proposed tag suggestion algorithm is explained in Section 3, experimental work and evaluation techniques are presented in Section 4, results are discussed in Section 5, and finally, Section 6 includes conclusions and possible future work.

2   Related Work

Automatic and semi-automatic annotation of photos has been widely studied throughout the years. Many studies in the field use text based, probabilistic and frequency oriented methods which all have their own restrictions.

Elliot and Ozsoyoglu describe a method for semi-automated semantic digital photo annotation in [9]. Related concepts, keywords, time and location information of a target photo are used for generating a set of related photos, and their tags are ranked according to a scoring function. Naaman *et al.* [1] use a context based approach for the annotation of persons in a photo. Their method requires time and space information when each photo is taken. Although, with wide usage of digital cameras, time information can easily be retrieved, space (i.e. physical location) information can only be obtained from a system with GPS support. Yan *et al.* [2] categorize tags in two categories: some of them are used in browsing and the others are used in tagging. The tags that are chosen for browsing purposes are not suggested to the users; only the remaining set of tags can actually be used for annotation purposes. To make this categorization, the method makes a frequency-based analysis and words that are widely used are chosen as good candidates for making efficient and effective browsing operations. The less frequently encountered words however, are seen to have discriminative properties so they are used in actual tagging.

There are not many work on annotation suggestion methods that use both keyword-based and visual feature based similarities. Wenyin *et al.* propose a progressive semi-automatic image annotation strategy that use keyword-based and content-based image retrieval and relevance feedbacks in [3]. They claim that manual annotation is a tedious but very accurate process, since tags are selected based on human determination of the semantic content of images. This strategy is used in *MiAlbum* system [4] and evaluations show that it is effective for annotating images in photo databases. The system annotates multiple photos when a query is given with the user's feedback. Similarly, Suh and Bederson describe their approach for efficient bulk annotations in [5,6], and they create meaningful image clusters for this purpose. However, an automatic approach is desired in our case, and we need to suggest tags for one photo. Therefore, annotation of groups of photos by the given strategies does not solve our annotation problem.

It is generally thought that computer vision techniques create a heavy overload and textual methods can be enough to provide good systems and thus they are not prefered to be used by researchers. But, though they are still developing, vision based methods are quite powerful and when combined with textual methods, very effective automated systems can be achieved. A good application of combined use of textual and visual techniques is proposed by Quack *et al.* in [11]. Objective of the work persented in [11] is to provide a system that automatically forms high quality image databases using the large-scale internet sources. They retrieve large numbers of raw data consisting of geotagged images togehter with their corresponding associated information(which include tags, title, description, time stampts etc.). They use textual, visual and spatial information to cluster these images. Then they classify their clusters into 'objects', which they define to be physical items on fixed locations, and 'events', special social occasions taken place at certain times. Using these specific classes in formed clusters, they associate images to wikipedia articles and check the validity of these associations. The final output of the system then becomes nicely organized groups of images with relevant enclopedia information attached to them.

We have observed that pure text-based approaches cannot provide perfect systems, as the visual content is totally independent from the textual content. Therefore the proposed method uses the advantages of both visual and textual techniques for obtaining high performance.

3   Tag Suggestion Algorithm

Our system is a stand-alone application that serves as an interface for uploading photos to Flickr. Main purpose of the system is to recommend tags to users as a photo is being uploaded, so that the probability of entering irrelevant tags to Flickr is reduced. To do this, user is required to provide initial tags with which the system retrieves related photos. Recommended tags are chosen among the distinct tags that come along with the set of related photos. While recommending a tag, visual similarities between a related photo and the photo to be uploaded are taken into account.

Figure 2 visually describes the proposed method. Algorithm steps are explained in further detail in the following subsections.

The method can be summarized in the following steps:

1. Obtain target photo and corresponding initial tags from user. Let I_t be the target photo to be uploaded, and $T_{init} = \{t_{init1}, t_{init2}\}$ be the initial tags for this photo.
2. Connect to Flickr server and fetch the first m relevant photos $I_R = \{I_1, ..., I_m\}$ (and their corresponding tags) $T(I_i)$ containing the given initial tags.

$$\forall I_i \in I_R, \ T_{init} \subset T(I_i) \tag{1}$$

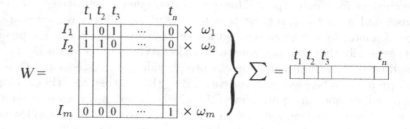

Fig. 1. Calculation of total weights W for each distinct tag in T

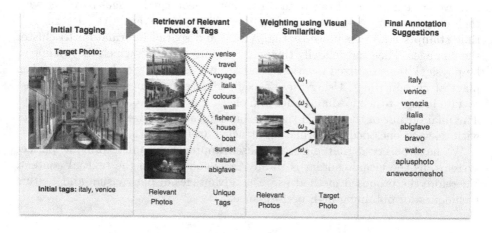

Fig. 2. Overview of the proposed method

3. Let $T = \{t_1, t_2, ..., t_n\}$ be the unique set of tags of all relevant photos, which contains n distinct tags.

$$\forall I_i \in I_R, \; T(I_i) \subset T \tag{2}$$

4. Extract visual feature f_{I_t} for the target photo, and f_{I_i} for all relevant photos.
5. Find the weight ω_i representing the visual similarity between target photo and the ith relevant photo I_i as Eqn 3.

$$\omega_i = \frac{1}{dist(f_{I_t}, f_{I_i})}, \; i \in \{1, ..., m\} \tag{3}$$

6. Generate a binary $m \times n$ matrix C, (where n is number of unique set of tags, m is the number of relevant photos). Set (i, j), if photo I_i contains tag t_j.

$$C_{ij} = 1 \Leftrightarrow t_j \in T(I_i) \tag{4}$$

7. Multiply each row i with the visual similarity ω_i, sum the columns to get a $1 \times n$ matrix W of tag weights as shown in Fig 1 where

$$W_i = \sum_{j=1}^{m} C_{ji} \times \omega_j \qquad (5)$$

8. Suggest tags in T according to their total weights W in non-decreasing order.

3.1   Initial Tagging and Retrieval of Related Photos from Flickr

In order for the system to be able to recommend tags, users need to provide a photo to be uploaded, together with a number of *initial tags*. These initial tags are the ones that need to be given by the user without any restrictions coming from the system. Although initial tags can be chosen freely, choosing the most descriptive tags as initial ones is encouraged for effective results.

The purpose of the set of initial tags is to retrieve a set of photos from Flickr that have a higher probability of being related to the target photo to be uploaded. When retrieved, recommended tags are chosen among the tags of this set of related photos. We define a *related photo* to be a photo that contains all of the initial tags provided by the user. In other words, the initial tags are AND'ed in order to obtain these photos from Flickr. As mentioned before, once identified, both the complete list of tags and the photos themselves are fetched from Flickr.

The main assumption of our recommendation system is that the tags initially given by a user has a high probability of being used in photos *similar* to that photo. Therefore, after retrieving the set of related photos and their corresponding tags, a new set containing distinct tags is formed. In the following steps, weights are given to these tags with respect to the similarity factor between the target photo and photo(s) that contain that tag. Finally, tags with higher weights are recommended to the user.

3.2   Visual Feature Extraction and Similarity Calculation

Finding visual similarities between our target photo and related photos that are retrieved from Flickr is a very crucial step for our method, therefore visual features are needed to be extracted from each image. There are a number of alternative approaches for selecting and implementing visual features. In our method two common and effective visual features, namely color histograms and interest points, have been implemented for evaluation.

For the color histograms, we considered two color spaces; the RGB and the HSV color spaces and found the results from HSV color space to be more effective. Using 8 bins for each band, we obtained a feature vector of length 24 for each image and we calculated the similarity between a pair of images by calculating the Euclidean distance between the feature vectors.

For the finding the interest points, the SIFT operator, [10], is used. From these interest points, similarities between image pairs are calculated by using the matching algorithm provided by Lowe [10]. The total number of match points between compared images are used as the similarity measure, and this value is normalized by the number of keypoints in target image.

3.3   Final Tag Suggestions

As mentioned before, a scoring function is applied to our list of candidate tags. We have used the visual features and their corresponding similarity measures separately to evaluate both approaches. However, in both cases, our scoring functions compute a weighted sum of the similarity values. After all candidate tags have been assigned a weight, tags with highest total weights are suggested to the user.

4   Experimental Work

To experiment on our work, we gathered 100 randomly chosen target photos from Flickr. Among these 100 photos there were some that were not suitable for our tests due to having too few number of tags or too specific tags (the necessity of these parameters will be clarified in section 4.3). Unsuitable photos were eliminated and performance measurements were made on the outputs of the remaining 66 target photos. In order to make a complete evaluation of the system, all tags of these set of target photos were analyzed and initial tags were chosen manually. As a design choice, size of the related photo set was selected as 100, thus about 7000 photos have been processed throughout the experiments. The following subsections describe experimenting environment, experiment results and evaluation methods are in further detail.

4.1   Experimental Environment: Flickr

As it has been mentioned before, proposed system is specifically designed to be used for the web site Flickr. With hundreds of millions of photos and over eight million users, Flickr is a rapidly developing web site that has a high potential of becoming a good source to be used by researchers working on social networking and content based image retrieval. Perhaps the most important reason for this rapid growth of attention is Flickr's emphasis on tagging. Through the use of tags, Flickr provides an image-browsing environment with various capabilities. As it is stated in Marlow *et al.*'s work in [7], Flickr has the following characteristic properties:

- *user-contributed* resources where users provide photos,
- *self-tagging* restrictions in which users can only tag the photos they have uploaded,
- *blind-tagging* behavior in terms of tagging support; tagging user cannot view other tags and the system does not suggest any possible tags to the user.

Tagging characteristics of Flickr are further discussed in [8]. According to the studies, although some photos contain more than 50 tags, statistics show that photos with 1 to 3 tags covers more than 60% of all photos. Most frequent category types of these tags are locations, objects, people, actions, and time.

We have implemented our system using Java programming language with the Flickr API[2], which is available for non-commercial use. Flickrj[3], a wrapper library for Flickr API, was used for querying the database. According to Flickr APIs Terms of Use agreement, after color features and invariant points are extracted from a photo, we do not cache Flickr's data.

4.2   Choosing Optimal Number of Initial Tags

For our experiments, we first examined the optimal number of initial tags to be specified. In our proposed method, users provide initial tags for the photo to be uploaded, and then we retrieve the contextually relevant photos in Flickr by using these initial tags. The system needs to retrieve about 100 relevant photos per target photo, so initial tags should be selected carefully. First of all initial tags should not be too specific as they would not return sufficient number of related photos. Second important factor is the number of initial tags to be used. When few number of general tags are chosen, we end up having thousands of relevant photos, of which only a small portion is used. On the other hand, as the number of initial tags increases, we get fewer and more specific photos. In this case, the number of relevant photos may not be enough for effective recommendations. From our studies we have found that using 2-3 initial tags gives the best results considering for the proposed method. The table of average number of relevant photos in Flickr for a given number of initial tags are given in Table 1.

Table 1. Average number of relevant photos for a given number of initial tags

Number of initial tags	Number of relevant photos
1	514044
2	4050
3	95
4	14
5	4

4.3   Evaluation Methods

There are various testing methods for correctly evaluating our method. We have considered the option to use already tagged photos in Flickr. In this approach we selected existing photos from Flickr, took a subset of their corresponding tags to be used as initial tags, and then compared the output of the system with original tags of each photo. For statistical analysis, we calculated precision and recall values. We compared original tag list with the annotation suggestions

---

[2] Flickr API, http://www.flickr.com/services/api/
[3] Flickrj, http://flickrj.sourceforge.net/

we gathered by using tag frequency, color features, and SIFT similarities. Tag frequency results can be used as a base line for comparing our results with only textual tag suggestion methods.

The evaluation metric accuracy is computed as the ratio of correctly suggested tags to the total number of suggested tags. Definition of accuracy A and overall accuracy A_{Avg} are formulated in Eqn 7, where ST is the suggested tags and T is the original tags.

$$A(I_i) = \frac{\mid T_i \cap ST(I_i) \mid}{\mid ST(I_i) \mid} \tag{6}$$

$$A_{Avg} = \frac{\sum_{i=1}^{n} A(I_i)}{n} \tag{7}$$

4.4 Experiment: Suggest-All-Tags

In our analysis, our first experiment was to *Suggest-All-Tags*, where we tried to suggest all the original tags of a photo by suggesting the same number of tags to the user. Accuracy is calculated among all original tags. For instance, if a Flickr photo in our test set has originally 22 tags, and we select 2 of them as initial tags, we suggest 20 tags to the user and try to make them match with the original tag list.

Figure 3 displays several different results in the form of recall vs. precision graphs we have obtained in our experiments. From these graphs we can see the change of performance of our visual similarities, namely color histograms and SIFT descriptors, as opposed to the performance of purely text-based tag frequency approach.

For photos of (a), (b), (c) and (d) results of color features produce relatively higher performance. As it can be visually observed, colors of photos (a), (b) and (d) are very significant and make them easier to be matched to other related photos. Colors of photo (c) are not as discriminant as the other three photos and thus the performance of frequency method is closer to color histogram method.

For photos of (g) and (h) results of all three approaches are approximately close to each other and their performances are low. SIFT features are better for scenes with specific objects. However, these photos are cluttered with objects; this explains the low performance. Moreover, as these photos do not contain distinctive colors and good illumination, color features also show low performance.

4.5 Experiment: Suggest-Top-5

Aim of the *Suggest-Top-5* experiment is to retrieve the relevant tags in the top 5 suggestions, which are the most important ones for the users.

Figure 4 shows the accuracies for 15 of the photos in the set of selected Flickr photos. These results represent more or less the overall results. We can say that our method slightly increases the results in most of the instances, has a better result in some of them (d, k, l, n and o), and results are not very good in a small number of photos (f and g).

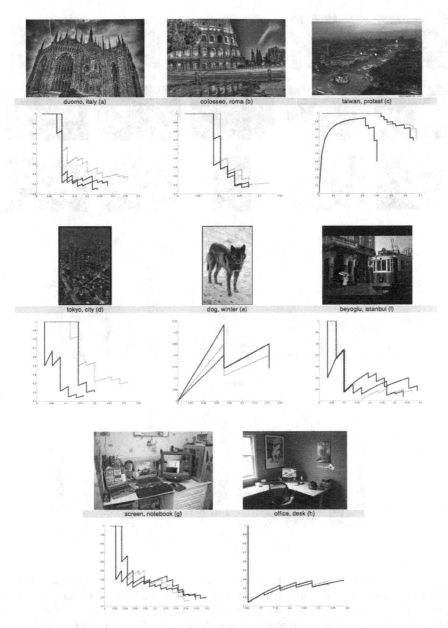

Fig. 3. Recall vs. Precision plots, their corresponding test photos together with their initial tags for Suggest-All-Tags experiment. Here, plots of color histogram method are drawn in gray, plots of SIFT similarity are drawn with a straight line, and plots of tag frequency are drawn with a dashed line. For the first four inputs (a-d), color similarity gives higher performance. Inputs (e) and (f) are exemplify high SIFT method performance. Remaining two inputs are examples for approximately close results from all three methods.

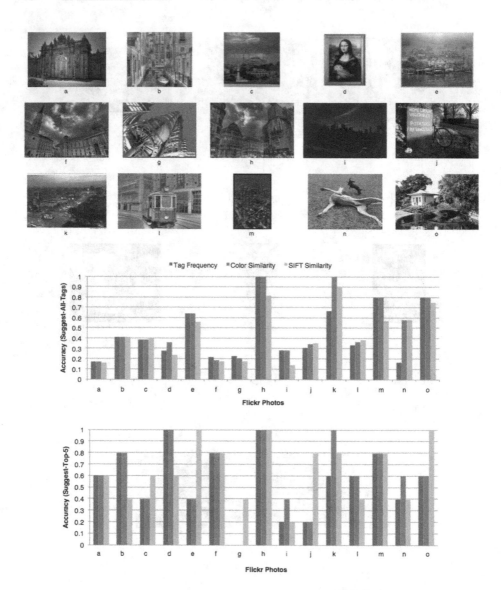

Fig. 4. Accuracy of all three methods tested on 15 given photos for Suggest-All-Tags and Suggest-Top-5 experiments. For both graphs, bars with lightest shade represent accuracy of SIFT similarity, bars with normal shade represent accuracy of Color similarity and bars with dark shade represent accuracy of tag frequency approach.

Table 2 represents the overall accuracies for both experiments. Results for color similarity are best among the three approaches. Suggest-Top-5 experiment has given significantly higher results in general because suggesting too many tags always reduces statistical performance results. Lowest accuracy is achieved when SIFT similarities are used.

Table 2. Avarage accuracy values of all three approaches for both Suggest-all-tags and Suggest-Top-5 experiments

Method	Accuracy for Suggest-all-tags	Accuracy for Suggest-Top-5
Tag Frequency	29%	52%
Color Similarity	31%	53%
SIFT Similarity	27%	46%

5   Discussion

Tagging is a very user-dependent process and validity checking of tags is difficult. Since we need to statistically evaluate tags suggested by our system, we have to define a basis for 'correct tags'. Normally, accepting user specified tags in Flickr as ground truth would be a reasonable approach. However, from our observations, we noticed that users do not properly tag their photos, and most of the tags are generally irrelevant to the image content. They have the tendency to add many commonly-used tags to make their photos popular.

Moreover, it is actually not truthful to state that, even for cases when original tags obtained from Flickr do not contain irrelevant tags, direct comparisons between a ground truth set and recommended tag set would give fully reliable results. Because for such systems, knowing that a certain tag has not been used by a user does not mean, that tag is irrelevant for that image. Different people may notice different aspects of a photo and it is not possible to represent all aspects of an image in words.

Due to these reasons, we here by claim that the statistical low performance we have observed in our experiments are not necessarily reflecting the truth. We have encountered numerous examples where our system suggested proper tags even though statistical results claimed them to be poor.

6   Conclusions and Future Work

In this paper, we proposed an automatic tag suggestion method which expands the tags of a photo by using both textual and visual information. Our motivation is to provide a system that enhances upload capabilities of a photo-sharing website so that users will easily select meaningful tags for their photos. The significance of our work is the approach of including visual information to the suggestion process. We have evaluated our system and showed that with current technologies, integrating visual features to automated systems do not add an unbearable overhead.

Automatic tagging systems have the potantial to be improved in many ways. As a future study, target photo can be examined in order to decide whether color similarity or keypoint similarity approach gives better annotation suggestions. Effective use of invariant keypoints in photos will enable our system to identify human-made objects and logos in a photo. Furthermore, developments in computer vision techniques should improve the performance of such systems where visual similarity is involved.

For improving our statistical results, as a future work, a user study can be prepared where a photo and a list of initial tags are given, and the users choose the other relevant tags. Results of such a study can be used as the ground truth for evaluating annotation suggestion methods and would provide more reliable results.

Acknowledgments

This research is partially supported by TÜBİTAK Career grant number 104E065 and grant number 104E077.

References

1. Naaman, M., Yeh, R.B., Garcia-Molina, H., Paepcke, A.: Leveraging context to resolve identity in photo albums. In: Proceedings of the 5th ACM/IEEE-CS Joint Conference on Digital Libraries, JCDL 2005, Denver, CO, USA, June 07 - 11, pp. 178–187. ACM, New York (2005)
2. Yan, R., Natsev, A., Campbell, M.: An efficient Manual Image Annotation Approach based on Tagging and Browsing. In: Proceedings of ACM International Multimedia Conference (2007)
3. Wenyin, L., Dumais, S., Sun, Y., Zhang, H., Czerwinski, M., Field, B.: Semi-automatic image annotation. In: Proc. of Interact: Conference on HCI, pp. 326–333 (July 2001)
4. Wenyin, L., Sun, Y., Zhang, H.: MiAlbum - a system for home photo managemet using the semi-automatic image annotation approach. In: Proceedings of the Eighth ACM international Conference on Multimedia, MULTIMEDIA 2000, Marina del Rey, California, United States, pp. 479–480. ACM, New York (2000)
5. Suh, B., Bederson, B.B.: Semi-Automatic Image Annotation Using Event and Torso Identification. Tech Report HCIL-2004-15, Computer Science Department, University of Maryland, College Park, MD
6. Suh, B., Bederson, B.B.: Semi-automatic photo annotation strategies using event based clustering and clothing based person recognition. Interact. Comput. 19(4), 524–544 (2007)
7. Marlow, C., Naaman, M., Boyd, D., Davis, M.: HT06, tagging paper, taxonomy, Flickr, academic article, to read. In: Proceedings of the Seventeenth Conference on Hypertext and Hypermedia, Odense, Denmark, August 22-25 (2006)
8. Sigurbjrnsson, B., van Zwol, R.: Flickr tag recommendation based on collective knowledge. In: Proceeding of the 17th international Conference on World Wide Web, WWW 2008, Beijing, China, April 21 - 25, pp. 327–336. ACM, New York (2008)
9. Elliott, B., Ozsoyoglu, Z.M.: A comparison of methods for semantic photo annotation suggestion. In: 22nd International International Symposium on Computer and Information Sciences, 2007. ISCIS 2007, November 7-9, pp. 1–6 (2007)

10. Lowe, D.: Distinctive image features from scale-invariant keypoints. International Journal of Computer Vision 60(2) (2004)
11. Quack, T., Leibe, B., Gool, L.V.: World-scale Mining of Objects and Events from Community Photo Collections. In: CIVR 2008, Niagara Falls, Canada, July 7-9 (2008)
12. Smeulders, A.W.M., Worring, M., Santini, S., Gupta, A., Jain, R.: Content based image retrieval at the end of the early years. IEEE Trans. on Pattern Analysis and Machine Intelligence 22(12) (2000)

Semantic Middleware to Enhance Multimedia Retrieval in a Broadcaster

Gorka Marcos[1], Petra Krämer[1], Arantza Illarramendi[2], Igor García[1],
and Julián Flórez[1]

[1] VICOMTech, Mikeletegi Pasealekua 57, Parque Tecnológico,
20009 Donostia-San Sebastián, Spain
gmarcos@vicomtech.org
[2] LSI Department, University of the Basque Country, Apdo. 649,
20080 Donostia-San Sebastián, Spain

Abstract. The digitalization of video and recent progress in semantic multimedia indexing and retrieval transform the workflow and tools involved in the information retrieval process of the broadcasters. To this end, we present in this paper a theoretical framework which addresses the semantic needs of this workflow from a semantic-centric view. Accordingly, we propose a pluggable middleware designed to provide the services covering the semantic needs spread all over the workflow in the system, including, the needs of independent software modules, of archivists, and of journalists. It is then shown how this can be employed in a real system to index and retrieve rushes material in a broadcaster.

1 Introduction

During the last decade the metadata lifecycle in the massive audiovisual content creation environments has undergone a significant development. The migration from tape archives to digital libraries accessible on the Intranet has changed the way how the metadata is generated and how the agents involved in the metadata workflow [1]. This, of course, opens new opportunities to exploit the content.

To date, most of Serb's premonitions [2] have come true. For instance, nowadays the annotations of the broadcasters' archives are not generated and managed only by the archivist. Consequently, the metadata related with the production and the content are not treated in a uniform way. Besides, despite the existence of different standards for the management of the metadata, most of the solutions in the broadcast industry are proprietary or customized solutions [3].

In such a context, let us consider the case of a broadcaster that acquires the Panasonic PS2 professional camera that embeds DMS-1[4] compliant annotations in a MXF (Multimedia eXchange Format) [5] container. This allows the cameramen to add metadata from the very beginning of the generation of the content such as coordinates, information about the device, date and location. However, once the memory cards arrive at the ingest department, what happens with this metadata? How is this metadata manipulated and enriched as the

D. Duke et al. (Eds.): SAMT 2008, LNCS 5392, pp. 74–88, 2008.

workflow of the content continues? When, how and who modifies this piece of information? What happens if the content and its metadata are exchanged with another organization? How do the current architectures semantically support the digital workflow?

Since is not an unique answer to these questions [6, 7], this paper aims at describing a platform independent semantic middleware that centralizes all the issues related with the knowledge and semantics of the organization. Furthermore, the middleware is easily plugable to any Multimedia Asset Management system. Therefore, all the processes that require exploiting the semantics can be adapted to the nature of the organization as they rely on the proposed middleware.

Focusing on the lifecycle of the content in a multimedia information retrieval system, a lot of work has been done for using the semantics of the content and its context in order to improve a concrete process of that workflow. Exhaustive literature has been presented dealing with the use of semantic techniques to improve the query processing and natural language processing [8, 9], query expansion [10, 11, 12, 13], query adaptation and federation [14, 15, 17, 18], information integration [15, 19, 20], results ranking [21, 22, 23] and information visualization [24, 25]. However, there are few reported initiatives that aim to tackle the semantic needs that arise during the workflow, from a centralized perspective.

Candela et al. [26] highlight the lack of standards in order to implement the mechanisms to access these semantic services. To solve this situation, they propose the Information Mediator Layer, whose main target is to make that information accessible in a unified way for the higher level services. The Intelligent Media Framework [27] integrates several components of a retrieval system relying on the existence of the Knowledge Content Objects to provide access to services for the storage of media, knowledge models and metadata relevant for the live staging process and providing services for the creation and management and delivery of intelligent media assets. Wei and Ngo [28] propose an architecture module which is designed to solve in a generic way the semantic needs of two main processes of a multimedia digital workflow: the analysis of the content and the mapping of the queries into the internal vocabulary. In [15], Kerschberg and Weishar address several issues related to the use of conceptual modeling to support services oriented, advanced information systems. They propose an "infomediation" layer in order to present (different views), to handle (intelligent thesaurus creation and management) and to gather (wrappers to internet, image and text analysis, ...) the information. In our previous work, the Meta Level [29], we describe a semantic middleware implementation deployed in the WIDE project [18]. It gathers the semantic information and functionalities of an information retrieval system in combination with multiple information sources. Meta Level is in charge of the semantic needs to carry out the query creation, analysis, mapping and federation, the results ranker and evaluation and the concept-based visualization of the domain and results. The work of Catells [21] is mainly focused on the semantic search in the Semantic Web. However, it employs an ontology based schema that integrates the semi-automatic annotation, search and retrieval of documents.

The above cited articles have something in common: they propose a layer or an architecture involved in the provision of at least more than one service that rely on some kind of semantic resources. However, those approaches either assume some requirements for the content that avoid its integration in current broadcasters workflows [21, 27], or they strengthen the accessing of the semantic information excluding the provision of the services from their approach [15, 26], or they do not tackle the provision of the services from a generic perspective [28].

Therefore, we present in this paper a semantic middleware which, using the information of the semantic resources of the system, centralizes the provision of all the semantic services needed for the retrieval process. The design of the middleware that we propose has been deeply influenced by the current information retrieval systems and workflow in a broadcaster. The objective of this design has been to enrich that workflow with new services that exploit the semantics of the domain. In addition, this design covers the semantic needs of that workflow. Furthermore, our middleware design has been tested through its deployment in a system where it provides full support for critical tasks like the automatic indexing of multimedia, including fuzziness techniques and the conduction of knowledge during the analysis process. Finally, our middlware is a transparent module for the remaining components. It is responsible for mapping between the different external information encapsulation formats and the internal format and terminology. According to this, our implementation embeds several intermediation parsers (DMS1 of MXF [4], the format for information exchange between modules...) to improve the scalability of the system.

The structure of this paper is the following. Section 2 presents our reference model for multimedia information retrieval. In Section 3, the architecture and target of the proposed middleware is described. Section 4 presents a deployment of the middleware in a search system. Therein, we describe the different services provided by the middleware. Finally, Section 5 concludes our work and outlines future work.

2   Reference Model for Multimedia Information Retrieval

Here, we present our model for multimedia retrieval which is based on a extension made by Larson on the Soergel reference model [30]. Figure 1 depicts our specialization of this model for the broadcast domain. The specialization consists in the following points:

- First of all, the model has been extended with the browsing line, in order to include in the information retrieval, the user experience during the browsing, navigation of the results and possible refinement of the query. We propose to include this line in the model to consider the semantic services provided by the middleware in those steps of the information retrieval process.
- Taking into account that our information retrieval analysis is carried out from the perspective of the workflow of a broadcaster, the components have been adapted to the terminology and idiosyncrasy of that domain.

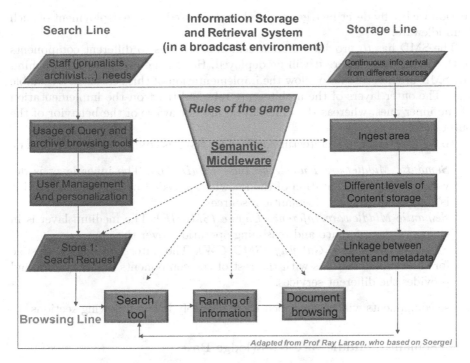

Fig. 1. Reference model for MIR in a broadcaster

- The content involved in the retrieval process is predominantly composed by audiovisual items. Therefore, the retrieval can be based either on the audiovisual features of the content (i.e. query by example techniques) or using the textual metadata gathered in the different steps of the workflow. The indexing of the material is more critical than in the case of text documents repositories. Each piece of the annotations composing the metadata of the asset will contribute significantly to the chances of the asset to be retrieved.
- Nowadays there are different metadata standards available defined both by the industry and the academia [6, 7]. However, it can be assumed that there is not a unique standard adopted by the broadcasters and that in many cases, different extended versions of various standards are used in different moments of the workflow.
- Considering that the limited amount of metadata available for those items and the diversity of the information, in a vision fully aligned with Styltsvig [31], our work supports the change from keyword-based to concept-based information retrieval utilizing ontologies as reference for concept definitions.

3   Generic Architecture of the Semantic Middleware

The Semantic Middleware (SMD) that we propose centralizes the semantic needs in a potential real deployment in an information retrieval system. Thus, in this section we briefly describe a generic architecture used for the deployment of such a middleware.

The SMD has to provide different semantic services to different components of the architecture where it will be deployed. Hence, the main objective behind the SMD architecture is to allow the implementation of this service in a scalable way. The outer layers of the architecture rely and trust on the implementation of the inner ones, whereas the inner layers are not aware of the behavior of the outer ones.

To achieve this, the architecture defined for SMD is composed by three layers:

- *Semantic Middleware Knowledge Base (SMD KB)*: The inner layer is devoted to the representation of the knowledge needed in the workflow and it is composed by several semantic resources.
- *Semantic Middleware Inference Engine (SMD IE)*: The medium layer is in charge of the inference and reasoning operations over the inner layer.
- *Semantic Middlware Gateway (SMD GW)*: The outer layer is responsible for the communications with the rest of the components of the system and provides the different services.

These components will be described more precisely in the following sections.

3.1   Semantic Middleware Knowledge Base

The SMD KB layer represents the "passive" knowledge modeled by the knowledge engineers and domain experts involved in the design of the information retrieval workflow and paradigm. From the information perspective, it is the key element of the middleware. Any piece of knowledge that must be used in order to solve the different needs of the system must be modeled here.

In order to implement the SMD KB the following criteria are crucial:

- There are different tools to gather the needed "knowledge": semantic repositories, populated ontologies, knowledge bases, syntactic grammars relational databases. According to the needs, expertise and idiosyncrasy of the environment where the SMD KB is going to be deployed, the optimum tool can be different. However, in any case, the tools have to guarantee the usability, maintenance and scalability of the information.
- The existence of an appropriate authoring tool for the experts involved in the creation and management of the information is decisive. Also the versioning control of the information is highly advisable.
- Together with the SMD KB, in a similar way to [15], a methodology and procedure for the correct generation and updating of the knowledge base has to be established. The different services that will be provided using this information, may impose different criteria for the definition of the relationships between the involved concepts. This has to be clear for the people involved in that task.
- Whenever possible, the usage of standards is very recommendable. In many sectors, the main concepts to be handled by the information retrieval are

agreed between the entities of the sector. These initiatives are mapped into standards that define common terminologies, exchange formats and common data models. The selected tool should be compliant with the inclusion of such initiatives.

3.2   Semantic Middleware Inference Engine

The SMD ID groups all the software items that we call *Processing Elements (PEs)*. PEs are related with the automatic extraction of knowledge out of the information stored in the SMD KB. Accordingly, this layer can include PEs that perform sequential query operations over a database in order to link several concepts or parse a sentence according to a syntactic grammar, to build a semantic graph out of a list of concepts, or to provide some feedback to a list of concepts according to some rules. The nature of these PEs is constrained by the following issues:

- They are highly dependent on the elements involved in the SMD KB.
- PEs perform an atomic semantic operation that could rely on other modules of the same layer.
- PEs are not context-aware. Therefore, the PEs do not need to know the objective of the action they are performing, in which part of the workflow they are invoked or even who is invoking them.
- There can be some intermediate PEs whose task is a merely mapping between some external available tools that handle the different elements of the SMD KB and some modules developed from scratch according to the needs of the concrete workflow.

3.3   Semantic Middleware Gateway

Finally, the SMD GW intermediates between the capacity of the inner layers of the SMD and the rest of the components of the retrieval system. Hence, the SMD GW is composed by different service providers or support processors that provide concrete semantic services hiding the knowledge represented in the SMD KB and the complexity of the PEs.

All the support processors in the SMD GW must be aware of the context of the service that they provide to the exterior and must use that information in order to invoke the different PEs of the SMD IE. The appropriate combination of the different invocations will lead the provision of the right service.

4   Semantic Middleware in a Multimedia Retrieval System

This section describes the real ongoing implementation of a SMD in a system developed in the context of the European project RUSHES[1] [32]. The objective of the project is to implement and validate a system for indexing, accessing and

---

[1] European project (FP6-045189), http://www.rushes-project.eu/

retrieving raw, unedited audio-visual footage known in broadcasting industry as "rushes". In order to accomplish this, many different technologies are required, including multimedia analysis, multimedia search, user interfaces as well as models for taxonomies and metadata. The project aims at testing and validating a proof of concept of the incoming semantic driven multimedia retrieval.

4.1   Metadata Model

The *Metadata Model (MDM)* is the SMD in the RUSHES system. Its architecture is based on a service-oriented architecture with loosely coupled services. Therein, the interfaces are usable without knowledge of the underlying implementation of the component exposing the service. The architecture defines a number of service domains, each of them represent some vital functionality of the RUSHES system. Functionalities exposed through services include storage, content processing, training of classification models, searching, and manual data annotation. The actual components implementing the service interfaces are hidden, and can be replaced by other components implementing the same service interface.

According to the architecture, as illustrated in Figure 2, the MDM constitutes a service domain. The MDM provides services to the following modules of the architecture: content capture and refinement module (CCR), the offline analysis component, the different user interfaces (through the query results and refinement module) and the search engine components. The different services provided to the different modules are described in section 4.2.

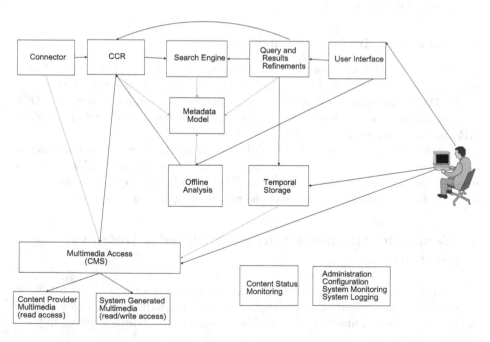

Fig. 2. General architecture of the RUSHES system

4.2   Semantic Services Provided by the Metadata Model

As previously mentioned, the MDM provides semantic services for the rest of the components during the different subprocesses that compose the RUSHES multimedia information retrieval workflow. Some of them are performed offline, before the user logs into the system, and others are invoked online. In the following the most important services are described and linked to the example of the Panasonic PS2 camera introduced in the first section of this paper.

Information Mediation. Before the user arrives, the video ingesting and analysis is performed. As a result, there are diverse annotations that belong to different semantic levels that are indexed by the search engine. During this process, the MDM carries out different parsing processes that should be taken between the external data schemas and the internal one defined in the system.

According to this the MDM is in charge of parsing the metadata provided by the PS2 camera (DMS-1 annotations) of the organization into the internal model. Beside this, once the analysis of the videos has been performed, which is explained in the next paragraph, the MDM is in charge of ensuring that these annotations arrive to the professional engine(s) that must handle them. In this case, this component is the FAST ESP engine, which is the key search-component in the media asset management of the broadcaster. In this case, the MDM parses the searchable annotations into a MEX (Multimedia Exhange Format) file. This MEX file, which is an XML document, is compliant with the schema defined in the search engine to index the annotations. This transforms the video into a retrievable document in the system. Beside this, the users are also able to use the manual annotation tool of the broadcaster to enrich those annotations.

Knowledge Conduction during the Multimedia Analysis Process. The CCR component provides an integrated environment for processing multimedia content such as image, audio and video. It provides an intuitive and efficient way for designing data flows from any content source to any content target. In the CCR component, a data flow is represented by an execution graph which consists of operators which are components performing some operation on data.

The set of operators invoked by the CCR is composed by expert modules (low level analysis operators, concept detectors and bayesian network classifiers) that work at different steps of the video analysis process. The information generated by some modules is needed by the remaining.

Some of them need additionally pieces of the semantic model in order to perform their analysis. During this process, the MDM is responsible for the persistency and availability of the intermediate information generated (i.e. low level features extracted, representative key-frames) and the semantic metadata obtained (i.e. list of faces recognized for each key-frame).

Furthermore, the MDM stores every piece of information generated for each video, preserving the semantic meaning of it by establishing its relation with the semantic entities gathered in its knowledge base.

Following the example, the video generated by the Panasonic cameras would be processed by the analysis algorithms of the broadcaster. Assuming that the analysis operators consist of a OCR text detector, a module which splits the video into shots and detects their key-frames, and a module to provide the probability of average number of faces in a shot, the MDM stores and relates the information provided by them. Additionally, the MDM ensures that the input needed by the face detector module is semantically equivalent to the information provided by the shot boundary detector.

Fuzzy Reasoning. Once the analysis of the video is finished, the MDM semantic repository is populated with all the information generated by the different operators. This information is related with the structure of the asset (number of tracks that compose the asset, main shots of the video and their representative keyframes) and with the content (number of faces present in each frame, type of audio, vegetation presence in a shot). These annotations, often linked to a confidence value, are inferred by the MDM in order to extract new knowledge.

In our example, the archivist, who is aware of the annotations provided by the stack of analysis modules, may employ the fuzzy inference engine in order to add the following rule: "A shot must be tagged with the word "DEMONSTRATION" and with a confidence value of 0,8 if the OCR detects the words "PEACE" or "NOT WAR" and if the shot contains more than 20 faces with a confidence value higher than 0,6".

Components of the User Interfaces. Within the RUSHES deployment, several interfaces are in charge of the interaction with the user in order to query the system, to retrieve and browse the results, and to annotate the assets. These interfaces implement interaction paradigms that employ tailored semantic services provided by the MDM. For instance, some of those services are: concept-based browsing, query recommendations based on the concepts typed by the user in the search, query parsing and enrichment, and support during the manual annotation.

Continuing with the example, when journalist searches a shot longer than 10 seconds with an average number of people bigger than 5, the semantics gathered by the MDM ("anything that contains a face contains a person") allows the the semantic mapping between the query related with the persons and the information about faces generated by the face detector.

4.3   Implementation of the Metadata Model

Here, we describe the implementation of the MDM that fulfills the requirements of the project and that follows the approach of the SMD presented in Section 3. The requirements are mainly derived from the specifications of the final users and from the general architecture defined in the project. The first group of requirements is mainly related with the functionality of the middleware (i.e. provided services, modeled domain, employed tools ...) whereas the second group of requirements has influenced the decisions related with the integration (i.e. language, fuzziness aspects, ...).

Metadata Model Knowledge Base. The MDM KB is composed of a set of interrelated OWL ontologies. The objective is to collect the work done in different initiatives according to the needs of the system. MDM KB is composed of the three ontologies:

- SMPTE 380: DMS-1 Ontology [4]: DMS-1 is the metadata schema of the MXF standards family. We migrated this schema into an OWL ontology to represent the structural information of the video where the semantic annotations must be attached to.
- MPEG-7 Detailed Audiovisual Profile (DAVP) [33]: In order to handle the information provided by the different analysis operators the MDM includes an OWL ontology of the DAVP profile.
- Domain Ontology for News Domain: The MDM includes an extended version of the LSCOM Lite ontology [34], which gathers the main concepts of the news videos domain according to the scope of the system.

For the edition of the model, the chosen language is OWL-DL. The edition is made offline by the experts with Protégé. The output is an OWL file (T-Box and A-Box) with all the information related to the domain.

In order to handle the uncertainty present in the multimedia analysis, OWL annotations are used. For instance, in Figure 3 an explanatory OWL fragment with such annotations is shown. This example expresses that "The key frame instance named as "10392" contains an instance of Face named as "Tony Blair" with a probability of 0,78". For the reasoning over this information, which is described later, a parsing between the T-Box and the fuzzy inference engine is done.

Metadata Model Inference Engine. The last version of the Jena API together with the FIRE fuzzy reasoner [35, 36] provided by NTUA have been used for the inference of the project. Regarding the FIRE reasoner, the input of the fuzziness inference is the A-Box and T-Box generated after the analysis. The MDM is responsible for the provision of the information needed by the Fuzzy reasoner: concepts, axioms and the instances and their probabilities.

Regarding the Jena API, it has been extended since, the methods implemented by this engine are mainly related with search and navigation of the concepts and instances of the model. The extension performed is due to the need to reduce the amount of time required by the API to browse and search the concepts of the model (T-Box), when the search criteria complexity is increased. For instance, inference methods have been implemented to enhance the Jena API in order to retrieve all the "intermediate" concepts and their subclasses that link the concepts A and B. This kind of knowledge extraction is needed, for example, for the generation of graphs that facilitate the generation of the query by the user.

Metadata Model Gateway. The MDM GW is composed by a set of webservices that expose the services mentioned in Section 4.2. The MDM GW is implemented as a standalone windows server and is able to attend parallel invocations from the different modules of the system. The server is stateless for all the services but the ones related with the analysis of the ingested videos. During that process, the server must keep the information of the different analysis

```xml
<?xml version="1.0"?>
<rdf:RDF
    xmlns:rdf="http://www.w3.org/1999/02/22-rdf-syntax-ns#"
    xmlns:xsd="http://www.w3.org/2001/XMLSchema#"
    xmlns:rdfs="http://www.w3.org/2000/01/rdf-schema#"
    xmlns:owl="http://www.w3.org/2002/07/owl#"
    xmlns="http://www.owl-ontologies.com/Ontology1196180634.owl#"
  xml:base="http://www.owl-ontologies.com/Ontology1196180634.owl">
  <owl:Ontology rdf:about=""/>
  <owl:Class rdf:ID="Face"/>
  <owl:Class rdf:ID="Video_Keyframe"/>
  <owl:ObjectProperty rdf:ID="contains">
    <rdfs:range rdf:resource="#Face"/>
    <rdfs:domain rdf:resource="#Video_Keyframe"/>
  </owl:ObjectProperty>
    <owl:DatatypeProperty rdf:ID="name">
    <rdfs:range rdf:resource="http://www.w3.org/2001/XMLSchema#string"/>
    <rdfs:domain>
      <owl:Class>
        <owl:unionOf rdf:parseType="Collection">
          <owl:Class rdf:about="#Video_Keyframe"/>
          <owl:Class rdf:about="#Face"/>
        </owl:unionOf>
      </owl:Class>
    </rdfs:domain>
  </owl:DatatypeProperty>
    <owl:DatatypeProperty rdf:ID="withAProbability">
    <rdfs:range rdf:resource="http://www.w3.org/2001/XMLSchema#float"/>
    <rdf:type rdf:resource="http://www.w3.org/2002/07/owl#AnnotationProperty"/>
  </owl:DatatypeProperty>
  <Video_Keyframe rdf:ID="keyframe_10392">
    <withAProbability rdf:datatype="http://www.w3.org/2001/XMLSchema#float"
    >0.78</withAProbability>
    <name xml:lang="en">10392</name>
    <contains>
      <Face rdf:ID="face_tonyBlair">
        <name xml:lang="es">Tony Blair</name>
      </Face>
    </contains>
  </Video_Keyframe>
</rdf:RDF>
```

Fig. 3. Annotation fragment with fuzziness information

operations performed on each video. In order to do that, a instance of part of the MDM GW is created to attend the requests related to each video.

In Figure 4, a partial view of this set of interfaces is shown. For example, the mdm2ccr represents the services provided by the middleware to the CCR module, in order to ensure the knowledge conduction during the video analysis process. In that view the functionalities are grouped according to the unit of information they are related with: the whole asset, a video segment, a cluster, a keyframe and so on. The methods available through these classes are used by

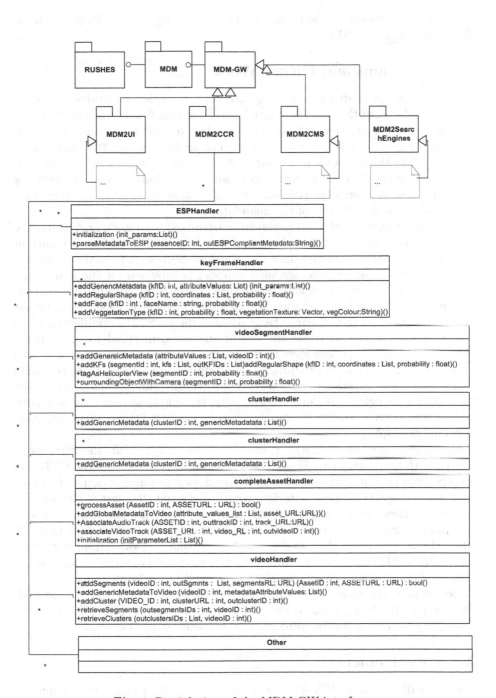

Fig. 4. Partial view of the MDM GW interfaces

the CCR module to store the information generated by the multimedia analysis algorithms in the MDM KB.

5   Conclusions and Future Work

First of all, we presented a generic architecture to implement customizable semantic middlewares that can be plugged into a multimedia information retrieval system to provide the different semantic services needed in the complex workflow, as the one present in a broadcaster. The main target of this middleware is to facilitate the adaptation of their systems to the new semantic needs generated by the employment of multimedia analysis techniques and the increase of the number of agents (journalists, camera manufacturers, industrial forums, ...) involved in the lifecycle of the metadata.

Secondly, we presented a real deployment where our middleware integrates the semantics distributed over the whole workflow in a unique model. The current services implemented are the ones described in Section 4.2. Initial metadata is obtained from the camera, semantic support is provided for the search and manual annotation processes, annotations are generated by the different analysis algorithms and linked with the annotations made by journalists and archivists using an annotation tool. During the analysis of the video, the middleware plays a central role for the semantic analysis of the video material by centralizing the management of the low level features together with the medium level and high level annotations. In this process, the middleware also becomes the main support for the fuzziness inference carried out over the annotations generated by the different analysis operators. Additionally, the middleware is able to handle DMS-1 annotations, thanks to the inclusion of the first DMS-1 OWL ontology in the bibliography.

Regarding the main activities planned for the next period, the first one is related with the evaluation. The MDM has been integrated with the CCR and the ESP components. This can be understood as a proof of concept of the architecture. However, the user interface components and the most of the analysis algorithms have not been finished and integrated. During the next months a first implementation of the whole RUSHES system will be available, which will allow the usage of real data and exhaustive testing and evaluation with final users. During this trials, the feasibility of the integration of a middleware as the presented one and the advantages of its usage will be performed.

Beside this, the increasing of the number of services provided by the middleware and the integration of the middleware knowledge base with external semantic resources accessible through Internet will be also targeted.

Acknowledgements

The RUSHES project (FP6-045189) is funded by the European Commission (FP6 program). Furthermore, we would like to thank the people from the Image, Video and Multimedia Systems Laboratory of National Technical University of Athens for the provision of the the Fuzzy-OWL reasoner.

References

1. Avilés, J.A.G., León, B., Sanders, K., Harrison, J.: Journalists at digital television newsrooms in Britain and Spain: workflow and multi-skilling in a competitive environment. Journalism Studies 5(1), 87–100 (2005)
2. Serb, K.: Towards the digital archive- A paradigm shift in exploiting mediaassets. In: Broadcasting Convention (1997)
3. Group, M.R.: Digital media asset management & workflow management in the broadcast industry: Survey & analysis. Technical report, Research And Markets (2004)
4. Smith, P.J.: Schirling: Metadata standards roundup. IEEE Multimedia 13, 84–88 (2006)
5. Haas, W., Mayer, H.: MPEG and its relevance for content-based multimedia retrieval. J. UCS 7(6), 530–547 (2001)
6. CHORUS: Workshop Annotations and Metadata models for Audiovisual/Multimedia Content IRT, Munich, November 21-22 (2007)
7. Cox, M., Tadic, L., Mulder, E.: Descriptive Metadata for Television. Focal Press (2006)
8. Zock, M.: Natural Language Generation: Snaphot of a fast evolving discipline (Recent Advances in Natural Language Processing). CNRS (September 2005)
9. Pérez, A.G., Galindo, M.S.B.: Ontologies and Natural Language Processing. In: Reflections on language use in the academic context. Universidad Politécnica de Madrid, pp. 121–161 (2005)
10. Smeaton, A.F., van Rijsbergen C.J.: The retrieval effects of query expansion on a feedback document retrieval system. The Computer Journal 26, 239–246 (1983)
11. Bhogal, J., Macfarlane, A., Smith, P.: A review of ontology based query expansion. Information Processing and Management: an International Journal 43(4), 866–886 (2007)
12. Bai, J., Nie, J.Y., Cao, G., Bouchard, H.: Using query contexts in information retrieval. In: SIGIR 2007: Proceedings of the 30th annual international ACM SIGIR conference on Research and development in information retrieval, pp. 15–22. ACM Press, New York (2007)
13. Song, M., Song, I.Y., Hu, X., Allen, R.B.: Integration of association rules and ontologies for semantic query expansion. Data Knowl. Eng. 63(1), 63–75 (2007)
14. Tzitzikas, Y., Constantopoulos, P., Spyratos, N.: Mediators over ontology-based information sources. In: WISE (1), pp. 31–40 (2001)
15. Kerschberg, L., Weishar, D.: Conceptual models and architectures for advanced information systems. Applied Intelligence 13(2), 149–164 (2000)
16. Garcia, R., Celma, O.: Semantic integration and retrieval of multimedia metadata. In: Proceedings of 4th International Semantic Web Conference. Knowledge Markup and Semantic Annotation Workshop (2005)
17. Arens, Y., Knoblock, C.A., Shen, W.M.: Query reformulation for dynamic information integration. J. Intell. Inf. Syst. 6(2-3), 99–130 (1996)
18. Smithers, T., Posada, J., Stork, A., Pianciamore, M., Ferreira, N., Grimm, S., Jimenez, I., Marca, S.D., Marcos, G., Mauri, M., Selvini, P., Sevilmis, N., Thelen, B., Zecchino, V.: Information management and knowledge sharing in wide. In: EWIMT (2004)
19. Kim, W., Seo, J.: Classifying schematic and data heterogeneity in multidatabase systems. Computer 24(12), 12–18 (1991)

20. Wache, H., Vögele, T., Visser, U., Stuckenschmidt, H., Schuster, G., Neumann, H., Hübner, S.: Ontology-based information integration: A Survey
21. Castells, P., Fernández, M., Vallet, D.: An adaptation of the vector-space model for ontology-based information retrieval. IEEE Trans. Knowl. Data Eng. 19, 261–272 (2007)
22. Shamsfard, M., Nematzadeh, A., Motiee, S.: Orank: An ontology based system for ranking documents. International Journal of Computer Science 1, 225–231 (2006)
23. Miller, G.: WordNet: A Lexical Database for English. Communications of the ACM 38(11), 39 (1995)
24. Faaborg, A.J.: A Goal-Oriented User Interface for Personalized Semantic Search. PhD thesis, B.A. Information Science Cornell University (2003)
25. Baeza-Yates, R.A., Ribeiro-Neto, B.A.: User Interaction and Visualization. In: Modern Information Retrieval, pp. 257–322. ACM Press / Addison-Wesley (1999)
26. Candela, L., Castelli, D., Pagano, P., Simi, M.: The digital library information mediator layer. In: Agosti, M., Thanos, C. (eds.) IRCDL, DELOS: a Network of Excellence on Digital Libraries, pp. 29–36 (2006)
27. Bürger, T.: An intelligent media framework for multimedia content. In: Proceedings of International Workshop on Semantic Web Annotations for Multimedia (SWAMM) (2006)
28. Wei, X.Y., Ngo, C.W.: Ontology-enriched semantic space for video search. In: MULTIMEDIA 2007: Proceedings of the 15th International Conference on Multimedia, pp. 981–990. ACM, New York (2007)
29. Marcos, G., Smithers, T., Jiménez, I., Toro, C.: Meta level: Enabler for semantic steered multimedia retrieval in an industrial design domain. Systems Science 2, 15–22 (2007)
30. Soergel, D.: Organizing Information: Principles of Data Base and Retrieval Systems, Orlando, Fl. Academic Press, London (1985)
31. Styltsvig, H.B.: Ontology-based Information Retrieval. PhD thesis, Computer Science Roskilde University, Denmark (2006)
32. Schreer, O., Ardeo, L.F., Sotiriou, D., Sadka, A., Izquierdo, E.: User requirements for multimedia indexing and retrieval of unedited audio-visual footage - rushes. In: Proc. of 9th Int. Workshop on Image Analysis for Multimedia Interactive Services (WIAMIS) (2008)
33. Troncy, R., Bailer, W., Hausenblas, M., Hofmair, P., Schlatte, R.: Enabling multimedia metadata interoperability by defining formal semantics of MPEG-7 profiles. In: Avrithis, Y., Kompatsiaris, Y., Staab, S., O'Connor, N.E. (eds.) SAMT 2006. LNCS, vol. 4306, pp. 41–55. Springer, Heidelberg (2006)
34. Neo, S.Y., Zhao, J., Kan, M.Y., Chua, T.S.: Video retrieval using high level features: Exploiting query matching and confidence-based weighting. In: Sundaram, H., Naphade, M., Smith, J.R., Rui, Y. (eds.) CIVR 2006. LNCS, vol. 4071, pp. 143–152. Springer, Heidelberg (2006)
35. Stoilos, G., Stamou, G., Pan, J., Tzouvaras, V., Horrocks, I.: Reasoning with very expressive fuzzy description logics. Journal of Artificial Intelligence Research 30, 273–320 (2007)
36. Simou, N., Kollias, S.: Fire: A fuzzy reasoning engine for impecise knowledge. In: K-Space PhD Workshop, K-Space PhD Students Workshop, Berlin, Germany, September 14 (2007)

Labelling Image Regions Using Wavelet Features and Spatial Prototypes⋆

Carsten Saathoff, Marcin Grzegorzek, and Steffen Staab

ISWeb – Information Systems and Semantic Web Research Group
Institute for Computer Science, University of Koblenz – Landau
{saathoff,marcin,staab}@uni-koblenz.de
http://isweb.uni-koblenz.de

Abstract. In this paper we present an approach for image region classification that combines low-level processing with high-level scene understanding. For the low-level training, predefined image concepts are statistically modelled using wavelet features extracted directly from image pixels. For classification, features of a given test region compared with these statistical models provide probabilistic evaluations for all possible image concepts. Maximising these values themselves already leads to a classification result (label). However, in our paper they are used as an input for the high-level approach exploiting explicitly represented spatial arrangements of labels, so called spatial prototypes. We formalise the problem using Fuzzy Constraint Satisfaction Problems and Linear Programming. They provide a model with explicit knowledge that is suitable to aid the task of region labelling. Experiments performed for nearly 6000 test image regions show that combining low-level and high-level image analysis increases the labelling accuracy significantly.

1   Introduction

It has been shown in various studies [1] that semantic access to multimedia content is desired by most users, regardless of whether they are professional or private users. An important field of research is automatic annotation of images, and specifically the automatic labelling of image regions [2]. Region-level annotations provide more detailed information about the image contents, allow for answering complex queries, and can be used to improve global classification accuracy [3].

Since exploiting solely low-level features often leads to unsatisfactory results, research towards using contextual and spatial features is a prominent research topic recently. A comprehensive study of using context for improving object recognition was carried out in [4, 5], showing the importance of contextual information. In [6] a survey of using spatial features for image region labelling based on graph models was performed and showed that spatial features improve

---

⋆ The research activity leading to this work has been supported by the European Commission under the contract FP6-027026-K-SPACE.

D. Duke et al. (Eds.): SAMT 2008, LNCS 5392, pp. 89–104, 2008.
ⓒ Springer-Verlag Berlin Heidelberg 2008

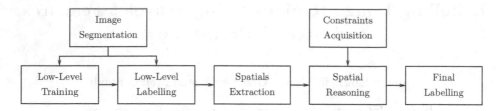

Fig. 1. Overall workflow of the approach for image region labelling

the labelling accuracy. In [7] an approach based on explicitly defined spatial constraints was introduced that employed genetic algorithms to compute a final labelling. We have also published results on exploiting explicitly represented spatial constraints for improving image labelling accuracy in [8].

The experiments in [8] indicate that our approach based on explicit representation of spatial context requires only a low amount of labelled examples for acquiring the explicit model. In this paper we conduct a new study focusing on the performance of our approach with different training set sizes. However, we combined the spatial reasoning part with a new low-level classification technique based on wavelet features, and provide a new formalisation of spatial constraints using binary integer programs. As we will show, the combination provides much better labelling accuracy with only few training examples.

The overall workflow of our method is depicted in Figure 1. The algorithm starts with the low-level training (Section 2.1). Here, all image concepts considered in our labelling task are modelled using feature vectors computed directly from image pixels. Instead of using MPEG-7 descriptors [9], we represent the image contents by wavelet features [10] and statistically model the concepts (e. g., sky, road, building, etc.) by density functions [11]. Subsequently, the low-level labelling is performed (Section 2.2). Both the training and labelling take advantage of automatic image segmentation algorithms. The low-level labelling results are used for further high-level processing, namely the extraction of spatial relations (Section 3) and spatial reasoning (Section 4). Here, we formalise the problem using Fuzzy Constraint Satisfaction Problems and Linear Programming. They provide a model with explicit knowledge that is suitable to aid the task of region labelling. Finally, the image region labels are provided by our algorithm. Results of experiments performed for nearly 6000 test image regions show that using the combination of low-level and high-level image analysis increases the labelling accuracy significantly (Section 5). This leads to some interesting conclusions presented in Section 6.

2   Content-Based Image Region Classification

In this section the low-level algorithm for content-based image region classification (labelling) is described, whereas the set of image concepts $\Omega = \{\Omega_1, \Omega_2, \ldots, \Omega_\kappa, \ldots, \Omega_{N_\Omega}\}$ is assumed to be a-priori known and constant. First, the

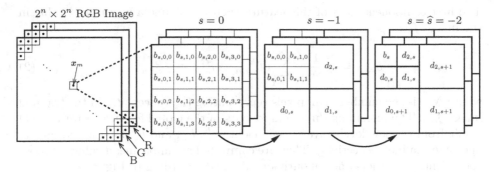

Fig. 2. Wavelet decomposition for a local neighbourhood of size 4×4 pixels done separately for the green, the red, and the blue channel. The final coefficients for the blue channel result from $b_{0,k,l}$ and have the following meaning: b_{-2} : low-pass horizontal and low-pass vertical, $d_{0,-2}$: low-pass horizontal and high-pass vertical, $d_{1,-2}$: high-pass horizontal and high-pass vertical, $d_{2,-2}$: high-pass horizontal and low-pass vertical.

statistical learning process is explained (Section 2.1). Second, the automatic labelling of image regions based on their contents is presented (Section 2.2).

2.1 Training of Image Concepts

In order to statistically train concepts (e. g., sky, road, building, etc.) based on image contents (pixel values), representative sets of example images for those concepts are required. The size of the training sets may vary, however, as you can see in Section 5, the performance of our image region classification algorithm depends on the number of training images. In order to reduce the amount of resources required to describe a large set of training images and to simplify the description, image contents are represented by feature vectors.

In our case RGB colour images are used for feature extraction. In order to calculate the vectors, a two-dimensional discrete signal decomposition with the wavelet transform [10] is applied for local neighbourhoods, whereas the Johnston 8-TAB wavelet is used as the basis function. A grid with size $\Delta r = 2^{|\widehat{s}|}$, where $\widehat{s}$ is the minimum multiresolution scale parameter[1] s, is overlaid on the image [12]. Figure 2 depicts this procedure for the case of colour scenes divided into local neighbourhoods of size 4×4 pixels. Further, the results of the low-pass filtering for all three colour channels (b_s^R, b_s^G, and b_s^B) in Figure 2 represented as b_s are taken into consideration for feature computation. Although the wavelet analysis is done for local neighbourhoods (see Figure 2), a training image should rather be described by a single global feature vector independent of the location in the image. For this reason the results of the local wavelet analysis are put together and their mean values are used for image description. Finally, each training image $f_{\kappa,i}$ obtains a global four-dimensional feature vector

$$c_{\kappa,i} = (c_{\kappa,i,1}, c_{\kappa,i,2}, c_{\kappa,i,3}, c_{\kappa,i,4})^{\mathrm{T}} \quad . \tag{1}$$

[1] Further decomposition of the signal with the wavelet transform is not possible.

The first component $c_{\kappa,i,1}$ of this feature vector is simple a mean pixel value in the image $\boldsymbol{f}_{\kappa,i}$

$$c_{\kappa,i,1} = \frac{1}{3 \cdot N_{\kappa,i}} \sum_{n=1}^{N_{\kappa,i}} (f_{\kappa,i,n}^R + f_{\kappa,i,n}^G + f_{\kappa,i,n}^B) \quad , \tag{2}$$

where $N_{\kappa,i}$ is the number of all pixels representing the concept Ω_κ in the training image $\boldsymbol{f}_{\kappa,i}$. The remaining three components of the global feature vector $\boldsymbol{c}_{\kappa,i}$ (1) result from the low-level wavelet coefficients b_s^R, b_s^G, and b_s^B for the red, green, and blue channel respectively. They are computed as simple mean values of those coefficients for all local neighbourhoods defined according to Figure 2

$$c_{\kappa,i,2} = \frac{1}{M_{\kappa,i}} \sum_{n=1}^{M_{\kappa,i}} b_{s,n}^R \quad , \tag{3}$$

$$c_{\kappa,i,3} = \frac{1}{M_{\kappa,i}} \sum_{n=1}^{M_{\kappa,i}} b_{s,n}^G \quad , \tag{4}$$

and

$$c_{\kappa,i,4} = \frac{1}{M_{\kappa,i}} \sum_{n=1}^{M_{\kappa,i}} b_{s,n}^B \quad , \tag{5}$$

where $M_{\kappa,i}$ is the number of all local neighbourhoods defined as in Figure 2 representing the concept Ω_κ in the training image $\boldsymbol{f}_{\kappa,i}$.

Since the number T_κ of training images $\boldsymbol{f}_{\kappa,i}$ for each concept Ω_κ is usually quite high[2], statistical modelling can be applied for training. It has been observed that the values of the feature vector components $c_{\kappa,i,n=1,\dots,4}$ behave regularly and can perfectly be modelled by normal density functions [13]. In order to do so, the mean values $\mu_{\kappa,n=1,\dots,4}$ and the standard deviations $\sigma_{\kappa,n=1,\dots,4}$ for the feature vector components $c_{\kappa,i,n=1,\dots,4}$ are computed in accordance to the well-known formulas

$$\mu_{\kappa,n} = \frac{1}{T_\kappa} \sum_{i=1}^{T_\kappa} c_{\kappa,i,n} \quad , \tag{6}$$

and

$$\sigma_{\kappa,n}^2 = \frac{1}{T_\kappa} \sum_{i=1}^{T_\kappa} (c_{\kappa,i,n} - \mu_{\kappa,n})^2 \quad . \tag{7}$$

Therefore, all concepts Ω_κ considered in the image region classification task are represented by a mean value vector

$$\boldsymbol{\mu}_\kappa = (\mu_{\kappa,1}, \mu_{\kappa,2}, \mu_{\kappa,3}, \mu_{\kappa,4})^{\mathrm{T}} \quad , \tag{8}$$

and a standard deviation vector

$$\boldsymbol{\sigma}_\kappa = (\sigma_{\kappa,1}, \sigma_{\kappa,2}, \sigma_{\kappa,3}, \sigma_{\kappa,4})^{\mathrm{T}} \tag{9}$$

after the training phase.

---

[2] In our experiments presented in Section 5 it varies from 50 to 400 in 7 steps.

2.2 Labelling of Image Regions

In order to classify image regions, first, a test image f is automatically segmented into test regions f_r. Then, each region found in the image f_r is described by a four-dimensional feature vector

$$c_r = (c_{r,1}, c_{r,2}, c_{r,3}, c_{r,4})^{\mathrm{T}} \quad . \tag{10}$$

This global feature vector is computed in exactly the same way as in the training phase (2, 3, 4, 5). The first component $c_{r,1}$ is a mean pixel value in the test region, while the remaining components $c_{r,n=2,\ldots,4}$ result from the wavelet analysis performed separately for the red, green, and blue channel of the image (see Figure 2). Now, for all possible concepts $\Omega = \{\Omega_1, \Omega_2, \ldots, \Omega_\kappa, \ldots, \Omega_{N_\Omega}\}$ trained as shown in Section 2.1, the comparison with the test region is performed. For this, density values $p_{\kappa,r,n=1,\ldots,4}$ for all feature vector elements $c_{r,n=1,\ldots,4}$ are computed using the trained mean (8) and standard deviation vectors (9) according to the definition of the Gaussian density function [11]

$$p_{\kappa,r,n} = p(c_{r,n}|\mu_{\kappa,n}, \sigma_{\kappa,n}) = \frac{1}{\sigma_{\kappa,n}\sqrt{2\pi}} \exp\left(\frac{(c_{r,n} - \mu_{\kappa,n})^2}{-2\sigma_{\kappa,n}^2}\right) \quad . \tag{11}$$

Assuming the statistical independency between the feature vector elements, the final evaluation of the test region represented by c_r and a hypothesis concept Ω_κ is computed with

$$p_{\kappa,r} = p(c_r|\mu_\kappa, \sigma_\kappa) = \prod_{n=1}^{4} p(c_{r,n}|\mu_{\kappa,n}, \sigma_{\kappa,n}) \quad . \tag{12}$$

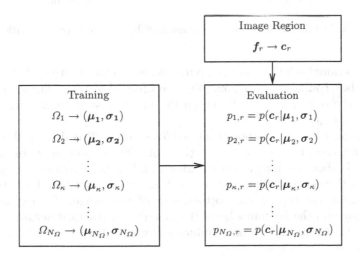

Fig. 3. The results of the training phase in form of mean vectors μ_κ and standard deviation vectors σ_κ for all concepts Ω_κ are compared with the image region to be classified f_r represented by c_r. As evaluation of this comparison density values $p_{\kappa,r}$ for all pairs "image region – hypothesis concept" are achieved.

Finally, the classification result $\Omega_{\widehat{\kappa}}$ (region label) is found by maximisation of the density value (12) over all possible concepts represented by their index κ

$$\widehat{\kappa} = \operatorname*{argmax}_{\kappa} p_{\kappa,r} = \operatorname*{argmax}_{\kappa} p(c_r | \boldsymbol{\mu}_{\kappa}, \boldsymbol{\sigma}_{\kappa}) \quad . \tag{13}$$

The correspondence between the training results and the image region to be classified is presented in Figure 3.

3   Spatial Relations Extraction

Within our region labelling procedure we consider four relative and two absolute spatial relations to model the spatial arrangements of the regions within an image. The relative spatial relations are *above-of, below-of, left-of, right-of*, and the absolute spatial relations are *above-all* and *below-all.*

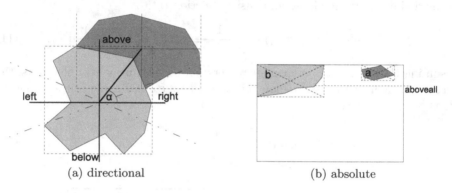

(a) directional   (b) absolute

Fig. 4. Definition of the a) directional and b) absolute spatial relations

The directional relations are computed based on the centres of the minimal bounding box containing a region. We have illustrated the definition of the directional relations in Fig. 4a. Based on the angle α we determine the relation between two regions.

Computing whether a region is *above-all* or *below-all* is done with three different approaches. First of all, we use the centre of the bounding box and check whether it is above (below) a certain threshold. Second, we take the region with the highest (lowest) bounding box centre. Finally, we also check for regions that "touch" *either* the top *or* the bottom edge of the image. If a region touches both the top and the bottom edge of the image, it is assigned neither of the two absolute regions, in order to not produce any contradictory constraints.

4   Spatial Reasoning Based on Constraints

The goal of the spatial reasoning step is to exploit background knowledge about the typical spatial arrangements of objects in images in order to improve the labelling accuracy compared to pure local, low-level feature-based approaches. As we will discuss in the following, the spatial background knowledge is automatically extracted from labelled examples, which we call the spatial prototypes. The knowledge consists of spatial constraint templates, which are explicitly represented spatial arrangements of concepts, possibly associated with a degree of confidence. We provide two formalisations of the problem, one based on Fuzzy Constraint Satisfaction Problems, which was already discussed in [8], and a new formalisation based on Linear Programming.

Our following discussions will be based on some fundamental formal definitions. Let L be the set of labels, and T the set of supported spatial relation types. An image is a tuple $I = \{S, R\}$, and S is the set of regions created by the segmentation. Each region $s_i \in S$ is associated with a membership function $\theta_i : L \to [0, 1]$ with $\theta_i = \{(l_k, p_{k,i})\}$, which associates each label with the probability provided by the low-level classification. Further, let $R = \{r_1, \ldots, r_k\}$ be the set of extracted spatial relations. An absolute spatial relation r is a region itself, i.e. $r \in S$, while relative spatial relations are pairs of regions, i.e. $r \in S^2$. Each spatial relation is associated with a type $t \in T$.

We will first discuss the acquisition of constraint templates from a set of spatial prototypes, and then describe the formalisation of the problem both using Fuzzy Constraint Satisfaction Problems and Binary Integer Programs.

4.1   Constraint Acquisition

Spatial constraint templates constitute the background knowledge in our approach. Manually defining these templates is a tedious task, specifically if the number of supported concepts and spatial relations becomes larger. We derive these templates from spatial prototypes, which are manually labelled images. We mine the prototypes using support and confidence as selection criteria, and come up with a set of templates representing typical spatial arrangements.

Let the set of prototypes be a set of images $P = \{I_1, \ldots, I_q\}$. For each region s_i of the images we assume that only one label exists, i.e. there is only one $l_i \in L$, such that $\theta_i(l_i) = 1$, and $\forall l_j \in L, l_j \neq l_i : \theta_i(l_j) = 0$. In the following we will say that l_i is the label associated with region s_i. We want to acquire one template for each spatial relation type $t \in T$. We denote the template as $\mathcal{T}_t$, and interpret each template as a fuzzy relation on the set of labels, i.e. $\mathcal{T}_t : L^n \to [0, 1]$. n equals 1 in the case of an absolute relation, and 2 for relative spatial relations.

In order to determine the membership degrees for each tuple of labels, we use the information present in our set of prototypes. For each label l we have to determine in what spatial relation to other labels it might be found. Therefore, for each spatial relation type $t \in T$, we consider the set of relations

$$R_{t\downarrow l} := \{r | \exists I = \{S, R\} \in P : \exists s_i \in S : \theta_i(l) = 1 \wedge r \in R_t\}, \qquad (14)$$

where $R_t \subseteq R$ denotes the set of relations with type t.

This set only contains relations from images depicting l, i.e. we limit both support and confidence to the *context* of label l. We then define $R_t^{l,l'} \subseteq R_{t\downarrow l}$

to be the set of relations between segments s, s' depicting l and l', respectively. Finally, we write $R_{t\downarrow l}^{*,l'} \subseteq R_{t\downarrow l}$ to denote all relations between an arbitrary region and a region depicting l'. The confidence of a label arrangement is then defined as

$$\gamma_t(l, l') = \frac{|R_t^{l,l'}|}{|R_{t\downarrow l}^{*,l'}|}, \tag{15}$$

and the support as

$$\sigma_t(l, l') = \frac{|R_t^{l,l'}|}{|R_{t\downarrow l}|}. \tag{16}$$

This definition can easily be modified for unary relations. In that case we are interested in the set R_t^l, which contains all absolute spatial relations on some region s that depicts label l. Further, we denote with $|l|$ the number of regions associated with label l. Support and confidence for absolute spatial relations can then be defined as

$$\gamma_t(l) = \frac{|R_t^l|}{|l|}, \tag{17}$$

and

$$\sigma_t(l) = \frac{|R_t^l|}{|R_{t\downarrow l}|}. \tag{18}$$

Finally, we have to define the template $\mathcal{T}_t$ for the spatial relation type t. We consider only two degrees of confidence for templates. We define $\mathcal{T}_t(l, l') = 1$ if we accept the pair, or $\mathcal{T}_t(l, l') = 0$ if we reject it. In order to determine whether we want to accept or reject a certain pair of labels for the relation t, we use two thresholds th_σ and th_γ, and define a template as

$$\mathcal{T}_t(l, l') = \begin{cases} 1 \text{ if } \sigma_t(l, l') > th_\sigma \text{ and } \gamma_t(l, l') > th_\gamma \\ 0 \text{ else} \end{cases} \tag{19}$$

For absolute spatial relations, the template is defined accordingly.

4.2   Spatial Reasoning with Fuzzy Constraint Satisfaction Problems

We transform the segmented and labelled image along with the spatial prototypes into a *Fuzzy Constraint Satisfaction Problem*. In the following, we will first introduce Fuzzy Constraint Satisfaction Problems as a formal model and then discuss the transformation. Our definition is based on [14] extended with *fuzzy domains*.

A Fuzzy Constraint Satisfaction Problem consists of an ordered set of fuzzy variables $V = \{v_1, \ldots, v_k\}$, each associated with the crisp domain $L = \{l_1, \ldots, l_n\}$ and the membership function $\mu_i : L \to [0, 1]$. The value $\mu_i(l), l \in L$ is called the degree of satisfaction of the variable for the assignment $v_i = l$. Further, we define a set of fuzzy constraints $C = \{c_1, \ldots, c_m\}$. Each constraint c_j is defined on a set of variables $v_1, \ldots, v_q \in V$, and we interpret a constraint as a fuzzy relation $c_j : L^q \to [0, 1]$. The value $c(l_1, \ldots, l_q)$, with $v_i = l_i$ is called the degree of satisfaction of the variable assignment $l_1, \ldots, l_q$ for the constraint c. In case that $c(l_1, \ldots, l_q) = 1$, we say that the constraint is fully satisfied, and if $c(l_1, \ldots, l_q) = 0$ we say it is fully violated. The purpose of fuzzy constraint reasoning is to obtain a variable assignment that is optimal with respect to the degrees of satisfaction of the variables and constraints. The quality of a solution is measured using a global evaluation function, which is called the *joint degree of satisfaction*.

We first define the joint degree of satisfaction of a variable, which determines the local satisfaction degree of the problem. Let $P = \{l_1, \ldots, l_k\}, k \leq |V|$ be a partial solution of the problem, with $v_i = l_i$. Let $C_i^+ \subseteq C$ be the set of the fully instantiated constraints on v_i. Further, let $\hat{c}$ stand for the degree of satisfaction of c given the current partial solution. Finally, let $C_i^- \subseteq C$ be the set of partially instantiated constraints on v_i. We then define the joint degree of satisfaction as $\text{dos}(v_i) := \frac{1}{\omega+1}(\frac{1}{|C_i^+|+|C_i^-|}(\sum_{c \in C_i^+} \hat{c} + |C_i^-|) + \omega \mu_i(l_i))$, in which ω is a weight used to control the influence of the degree of satisfaction of the variable assignment on the joint degree. In this definition we overestimate the degree of satisfaction of partially instantiated constraints to 1.

We now define the joint degree of satisfaction for a complete Fuzzy Constraint Satisfaction Problem. Let $J := \{\text{dos}(v_{i_1}), \ldots, \text{dos}(v_{i_n})\}$ be an ordered multiset of joint degrees of satisfaction for each variable in V, with $\forall v_{i_k}, v_{i_l} \in V, k < l :$ $\text{dos}(v_{i_k}) \leq \text{dos}(v_{i_l})$. The joint degree of satisfaction of a variable that is not yet assigned a value is overestimated to 1. We can now define a lexicographic order $>_L$ on the multisets. Let $J = \{\gamma_1, \ldots, \gamma_k\}, J' = \{\delta_1, \ldots, \delta_k\}$ be multisets. Then $J >_L J'$, iff $\exists i \leq k : \forall j < i : \gamma_j = \delta_j$ and $\gamma_i > \delta_i$. If we have two (partial) solutions P, Q to a Fuzzy Constraint Satisfaction Problem with according joint degree of satisfactions J_P, J_Q, solution P is better than Q, iff $J_P >_L J_Q$.

Now, we can transform an initially labelled image into a Fuzzy Constraint Satisfaction Problem using the following algorithm.

1. For each region $s_i \in S$ create a variable v_i on L with $\mu_i := \theta_i$.
2. For each region $s_i \in S$ and for each spatial relation r of type *type* defined on s_i and further segments $s_1, \ldots, s_k$ create a constraint c on $v_i, v_1, \ldots, v_k$ with $c := p$, where $p \in P$ is a spatial prototype of type *type*.

The resulting Fuzzy Constraint Satisfaction Problem can efficiently be solved using algorithms like branch and bound, as was also discussed in [14].

4.3   Spatial Reasoning with Linear Programming

We will show in the following how to formalise image labelling with spatial constraints as a linear program. We will first introduce Linear Programming as a formal model, and then discuss how to represent the image labelling problem as a linear program.

A linear program has the standard form

$$\begin{aligned} \text{minimize} \quad & Z = \mathbf{c}^T \mathbf{x} \\ \text{subject to} \quad & \mathbf{A}\mathbf{x} = \mathbf{b} \\ & \mathbf{x} \geq 0 \end{aligned} \tag{20}$$

where $\mathbf{c}^T$ is a row vector of so-called objective coefficients, $\mathbf{x}$ is a vector of control variables, $\mathbf{A}$ is a matrix of constraint coefficients, and $\mathbf{b}$ a vector of row bounds. Efficient solving techniques exist for linear programs, e.g. the *Simplex Method*. Goal of the solving process is to find a set of assignments to the variables in $\mathbf{x}$ with a minimal evaluation score Z that satisfy all the constraints. In general, most non-standard representations of a linear program can be transformed into this standard representation. In this paper we will consider binary integer programs of the form

$$\text{maximize} \quad Z = \mathbf{c}^T\mathbf{x}$$
$$\text{subject to } \mathbf{Ax} = \mathbf{b} \quad\quad\quad (21)$$
$$\mathbf{x} \in \{0,1\}$$

where $\mathbf{x} \in \{0,1\}$ means that each element of $\mathbf{x}$ can either be 0 or 1.

In order to represent the image labelling problem as a linear program, we create a set of linear constraints from each spatial relation in the image, and determine the objective coefficients based on the hypotheses sets and the constraint templates.

Given an image $I = \{S, R\}$, let $O_i \subseteq R$ be the set of outgoing relations for region $s_i \in S$, i.e. $O_i = \{r \in R | \exists s \in S, s \neq s_i : r = (s_i, s)\}$. Accordingly, $E_i \subseteq R$ is the set of incoming spatial relations for a region s_i, i.e. $E_i = \{r \in R | \exists s \in S, s \neq s_i : r = (s, s_i)\}$. For each spatial relation we need to create a set of control variables according to the following scheme. Let $r = (s_i, s_j)$ be the relation. Then, for each possible pair of label assignments to the regions, we create a variable c_{itj}^{ko}, representing the possible assignment of l_k to s_i and l_o to s_j with respect to the relation r with type $t \in T$. Each c_{itj}^{ko} is an integer variable and $c_{itj}^{ko} = 1$ represents the assignments $s_i = l_k$ and $s_j = l_o$, while $c_{itj}^{ko} = 0$ means that these assignments are not made. Since every such variable represents exactly one assignment of labels to the involved regions, and only one label might be assigned to a region in the final solution, we have to add this restriction as linear constraints. The constraints are formalised as

$$\forall r \in R : r = (s_i, s_j) \in R \rightarrow \sum_{l_k \in L}\sum_{l_o \in L} c_{itj}^{ko} = 1. \quad\quad\quad (22)$$

These constraints assure that there is only one pair of labels assigned to a pair of regions per spatial relation, but it does not guarantee, that for all relations involving a specific region the same label is chosen for the region. In effect, this means that there could be two variables c_{itj}^{ko} and $c_{it'j'}^{k'o'}$ both being set to 1, which would result in both k and k' assigned to s_i. Since our solution requires that there is only one label assigned to a region, we have to add constraints that "link" the variables accordingly.

We basically require that either all variables assign a label l_k to a region s_i, or none. This can be accomplished by linking pairs of relations. We first have to link the outgoing relations O_i, then the incoming ones E_i, and finally link one of the outgoing relations to one of the incoming ones. This system of linear constraints will ensure that only one label is assigned to the region in the final solution.

We will start by defining the constraints for the outgoing relations. We take one arbitrary relation $r_O \in O_i$ and then create constraints for all $r \in O_i \setminus r_O$. Let $r_O = (s_i, s_j)$ with type t_O, and $r = (s_i, s_{j'})$ with type t be the two relations to be linked. Then, the constraints are

$$\forall l_k \in L : \sum_{l_o \in L} c_{it_O j}^{ko} - \sum_{l'_o \in L} c_{itj'}^{ko'} = 0. \tag{23}$$

The first sum can either take the value 0 if l_k is not assigned to s_i by the relation r, or one if it is assigned. Equation (22) ensures that only one of the c_{itj}^{ko} is set to 1. Basically, the same applies for the second sum. Since both are subtracted and the whole expression has to evaluate to 0, either both equal 1 or both equal 0 and subsequently, if one of the relations assigns l_k to s_i, the others have to do the same. We can define the constraints for the incoming relations accordingly. Let $r_E \in E_i$, $r_E = (s_j, s_i)$ with type t_E be an arbitrarily chosen incoming relation. For each $r \in E_i \setminus r_E$ with $r = (s_{j'}, s_i)$ and type t create constraints

$$\forall l_k \in L : \sum_{l_o \in L} c_{jt_E i}^{ok} - \sum_{l'_o \in L} c_{j'ti}^{o'k} = 0. \tag{24}$$

Finally we have to link the outgoing to the incoming relations. Since the same label assignment is already enforced within those two types of relations, we only have to link r_O and r_E, using the following set of constraints:

$$\forall l_k \in L : \sum_{l_o \in L} c_{it_O j}^{ko} - \sum_{l'_o \in L} c_{j't_E i}^{o'k} = 0 \tag{25}$$

Absolute relations are formalised accordingly. Let $A_i \subseteq R$ be the set of absolute relations defined on s_i. For each $r \in A$ of type t we define a set of control variables c_{it}^k, $\forall l_k \in L$. The constraint enforcing only one label assignment is defined as

$$\sum_{l_k \in L} c_{it}^k = 1, \tag{26}$$

and we link it to all remaining absolute relations r' on the region s_i with

$$\forall l_k \in L : c_{it}^k - c_{it'}^k = 0. \tag{27}$$

Further we have to link the absolute relation to the relative relations. Again, linking one of the relations is sufficient, and therefore we choose either the relation r_O or r_E and an arbitrary absolute relation a:

$$\forall l_k \in L : c_{it_a}^k - \sum_{l_o \in L} c_{it_O j}^{ko} = 1 \tag{28}$$

Eventually, let t_r and t_a refer to the type of the relative relation r and the absolute relation a, respectively, then the objective function is defined as

$$\sum_{r=(s_i,s_j)} \sum_{l_k \in L} \sum_{l_o \in L} \min(\theta_i(l_k), \theta_j(l_o)) * \mathcal{T}_{t_r}(l_k, l_o) * c_{it_r j}^{ko} + \sum_{a=s_i} \sum_{l_k \in L} \theta_i(l_k) * \mathcal{T}_{t_a}(l_k) * c_{it_a}^k.$$
$$\tag{29}$$

For label pairs that violate our background knowledge, the product in the sum will evaluate to 0, while for pairs satisfying our background knowledge, we take the minimum degree of confidence from the hypotheses sets. Therefore we reward label assignments that satisfy the background knowledge and that involve labels with a high confidence score provided during the classification step.

5   Experiments and Results

Since we are interested in providing good labelling performance with only few training examples, we conducted an evaluation of our approach with varying training set sizes on a dataset of over 900 images with region-level annotations. The dataset was divided into a test and a number of training sets of different sizes. We then carried out a number of experiments on the largest training set to determine a set of spatial relation types and acquisition parameters used for the final evaluation. The differences achieved with variations in the parameters or spatial relations are only minor, so that we do not expect a large impact on the final results if the parameters are changed. We will continue by describing the image database first, and then give the results of our experiments both for the pure low-level approach, and the combination with the spatial reasoning. In the end we will discuss the lessons learned.

5.1   Image Database

The dataset consists of 923 images depicting outdoor scenes ranging from beach images over mountain views and cityscape images. An overview is provided in Figure 5. We chose a set of 10 labels that were prominent in the images and where a correct segmentation was feasible. The labels are *building, foliage, mountain, person, road, sailing-boat, sand, sea, sky, snow*. Additionally, we used the label *unknown* for regions where we could not decide on a definite label.

Fig. 5. Overview of the image database used for the experiments

For preparing the groundtruth, all images were segmented by an automatic segmentation algorithm that was available from a research project [15], and the resulting regions were labelled using exactly one of the labels. We always used the dominant concept depicted in a region, i.e. the concept covering the largest part of the region, and labelled the region with the according label. Regions without a dominant concept, or regions depicting an unsupported concept, were assigned the *unknown* label.

In total the dataset contained 6258 regions, of which 568 were labelled with the unknown label. This resulted in a dataset of 5690 regions labelled with a supported concept. 3312 were used for evaluation, and in the largest data set we used 2778 for training.

5.2   Experimental Results

The goal of our experiments was to determine the performance of our approach with varying training-set sizes. For that a series of experiments was performed in order to determine the influence of different parameters and features on the overall performance using the largest training set. We fixed those parameters and then performed the experiments using different training set sizes. The final setup consisted of the spatial relations discussed in Section 3, and using the thresholds $\sigma = 0.001$ and $\gamma = 0.2$ for both relative and absolute spatial relations. As one can see from the final thresholds, filtering on support is not feasible, but confidence provides a good quality estimation for spatial constraint templates.

For each approach, i.e. pure low-level classification, spatial reasoning using Fuzzy Constraint Satisfaction Problems, and spatial reasoning using Binary Integer Programs, we measured *precision (p)*, *recall (r)* and the *classification rate (c)*. Further we computed the *F-Measure (f)*. In Table 1 the average for each of these measures is given.

One can clearly see, that both spatial reasoning approaches improve the pure low-level classification results. This observation is fully consistent with earlier findings [8], and also other studies that were performed [6,4,7]. It is also obvious that the binary linear programming approach outperforms the fuzzy constraint satisfaction approach. This is probably due to the different objective functions

Table 1. Overall results for the three approaches

set size	Low-Level				FCSP				BIP			
	p	r	f	c	p	r	f	c	p	r	f	c
50	.63	.65	.57	.60	.65	.64	.62	.67	.77	.75	.73	.75
100	**.70**	.67	**.65**	**.69**	.67	**.67**	.65	.70	.78	**.77**	.75	.80
150	.67	.63	.61	.66	.66	.64	.63	.69	.74	.71	.70	.75
200	.69	.65	.63	.67	.67	.64	.64	.68	.80	.75	.76	.80
250	.69	.64	.60	.66	.69	.66	.65	**.70**	.78	.73	.72	.77
300	.68	.63	.61	.66	.68	.65	.64	.69	**.82**	**.77**	**.78**	**.82**
350	.63	**.68**	.61	.66	**.70**	.66	**.66**	.70	.80	.75	.76	.80
400	.68	.63	.61	.66	.69	.66	.65	**.70**	.80	.75	.75	.79

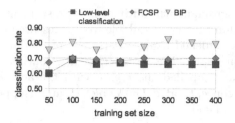

Fig. 6. Development of the classification rate with different training-set sizes

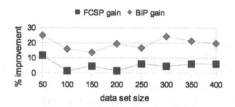

Fig. 7. The improvement over the low-level classification achieved with increasing number of training examples

used. The Lexicographic order is still a rather coarse estimation of the overall labelling accuracy, while the objective function for the binary integer program well integrates the two important properties, i.e. satisfying spatial constraints and a high probability score from the classifier.

In Figure 6 we have visualised the performance development with increasing data set sizes. Against our initial assumptions, we do not see a steady increase in performance, but already with 100 training examples nearly the best performance, which stays pretty stable for the rest of the experiments. One can also see that incorporating spatial context provides the largest performance increase for the smallest training set, which is an indicator that a good set of constraint templates is already acquired with only 50 prototype images. We have summarised the classification rate improvement achieved with the spatial reasoning in Figure 7.

5.3   Lessons Learned

The improvement is most significant for the training set with only 50 images. For this set, the low-level classification rate is worst, but the large improvement indicates that the background knowledge already provides a good model after acquisition from only 50 examples. The best overall classification rate is achieved with the binary integer programming approach on the data set with 300 training images. However, the classification rate with 100 training examples is nearly the same, which indicates that 100 training examples are a good size for training a well performing classifier.

In general, the experiments show that the combination of the statistical training of low-level classifiers with an explicit spatial context model based on binary integer programming provides a good foundation for labelling of image regions with only few training examples.

Further, our experiments also revealed that solving this kind of problem is much more efficient using binary integer programs. In average, the binary integer programming approach requires 1.1 seconds for one image, with a maximum value of 41 seconds and a minimum of only 6 ms. The fuzzy constraint reasoning, however, takes several hours for a few images, while in average it takes around 40 seconds. So, for the FCSP the runtime is much less predictable and also much higher in the average case. So, especially for real applications the binary integer programming approach clearly seems preferable.

6 Conclusions

In this paper we have introduced a novel combination of a statistical method for training and recognising concepts in image regions, integrated with an explicit model of spatial context. We have proposed two ways of formalising explicit knowledge about spatial context, one based on Fuzzy Constraint Satisfaction Problems, that was already presented in [8], and a new one based on Binary Integer Programming.

Our results show that the combination of both approaches results in a good classification rate compared to results in the literature. We have further evaluated how the classification rate develops with an increasing number of training examples. Surprisingly, nearly the best performance was already reached with only 100 training images. But also with 50 training images (approx. 344 regions) the combined approach provided a reasonable classification rate.

We are going to continue this work in the future introducing some improvements. For instance, new low-level features combining the texture and shape information will be applied for image content description.

References

1. Hollink, L., Schreiber, T.A., Wielinga, B.J., Worring, M.: Classification of user image descriptions. International Journal of Human-Computer Studies 61(5) (2004)
2. Barnard, K., Fan, Q., Swaminathan, R., Hoogs, A., Collins, R., Rondot, P., Kaufhold, J.: Evaluation of localized semantics: data, methodology, and experiments. International Journal of Computer Vision 77, 127–199 (2008)
3. Fan, J., Gao, Y., Luo, H.: Multi-level annotation of natural scenes using dominant image components and semantic concepts. In: Proc. of ACM Multimedia 2004, pp. 540–547. ACM, New York (2004)
4. Torralba, A.: Contextual priming for object detection. Int. J. Comput. Vision 53(2), 169–191 (2003)
5. Grzegorzek, M., Izquierdo, E.: Statistical 3d object classification and localization with context modeling. In: Domanski, M., Stasinski, R., Bartkowiak, M. (eds.) 15th European Signal Processing Conference, Poznan, Poland, PTETiS, Poznan, pp. 1585–1589 (2007)

6. Yuan, J., Li, J., Zhang, B.: Exploiting spatial context constraints for automatic image region annotation. In: Proc. of ACM Multimedia 2007, pp. 595–604. ACM, New York (2007)
7. Panagi, P., Dasiopoulou, S., Papadopoulos, T.G., Kompatsiaris, Strintzis, M.G.: A genetic algorithm approach to ontology-driven semantic image analysis. In: Proc. of VIE 2006, pp. 132–137 (2006)
8. Saathoff, C., Staab, S.: Exploiting spatial context in image region labelling using fuzzy constraint reasoning. In: WIAMIS: Ninth International Workshop on Image Analysis for Multimedia Interactive Services (2008)
9. Manjunath, B.S., Salembier, P., Sikora, T.: Introduction to MPEG-7 - Multimedia Content Description Interface. John Willey & Sons Ltd., Chichester (2002)
10. Mallat, S.: A theory for multiresolution signal decomposition: The wavelet representation. IEEE Transactions on Pattern Analysis and Machine Intelligence 11(7), 674–693 (1989)
11. Webb, A.R.: Statistical Pattern Recognition. John Wiley & Sons Ltd., Chichester (2002)
12. Grzegorzek, M., Reinhold, M., Niemann, H.: Feature extraction with wavelet transformation for statistical object recognition. In: Kurzynski, M., Puchala, E., Wozniak, M., Zolnierek, A. (eds.) 4th International Conference on Computer Recognition Systems, Rydzyna, Poland, pp. 161–168. Springer, Heidelberg (2005)
13. Grzegorzek, M.: Appearance-Based Statistical Object Recognition Including Color and Context Modeling. Logos Verlag, Berlin (2007)
14. Ruttkay, Z.: Fuzzy constraint satisfaction. In: Proc. of Fuzzy Systems 1994, vol. 2, pp. 1263–1268 (1994)
15. Dasiopoulou, S., Heinecke, J., Saathoff, C., Strintzis, M.G.: Multimedia reasoning with natural language support. In: Proc. of ICSC 2007, pp. 413–420 (2007)

Performing Content-Based Retrieval of Humans Using Gait Biometrics

Sina Samangooei and Mark S. Nixon

School of Electronics and Computer Science, Southampton University, Southampton,
SO17 1BJ, United Kingdom
{ss06r,msn}@ecs.soton.ac.uk

Abstract. In order to analyse surveillance video, we need to efficiently
explore large datasets containing videos of walking humans. At surveil
lance-image resolution, the human walk (their gait) can be determined
automatically, and more readily than other features such as the face.
Effective analysis of such data relies on retrieval of video data which has
been enriched using semantic annotations. A manual annotation process
is time-consuming and prone to error due to subject bias. We explore the
content-based retrieval of videos containing walking subjects, using semantic queries. We evaluate current biometric research using gait, unique
in its effectiveness at recognising people at a distance. We introduce
a set of semantic traits discernible *by humans* at a distance, outlining
their psychological validity. Working under the premise that similarity
of the chosen gait signature implies similarity of certain semantic traits
we perform a set of semantic retrieval experiments using popular latent
semantic analysis techniques from the information retrieval community.

1 Introduction

In 2006 it was reported that around 4 million CCTV cameras were installed
in the UK[4]. This results in 1Mb of video data per second per camera, using relatively conservative estimates[1] . Analysis of this huge volume of data
has motivated the development of a host of interesting automated techniques,
as summarised in[7][16], whose aim is to facilitate effective use of these large
quantities of surveillance data. Most techniques primarily concentrate on the
description of human behaviour and activities. Some approaches concentrate on
low level action features, such as trajectory and direction, whilst others include
detection of more complex concepts such as actor goals and scenario detection.
Efforts have also been developed which analyse non human elements including
automatic detection of exits and entrances, vehicle monitoring, etc.

Efficient use of large collections of images and videos by humans, such as
CCTV footage, can be achieved more readily if media items are meaningfully
semantically transcoded or *annotated*. Semantic and natural language description has been discussed [16] [41] as an open area of interest in surveillance. This

---

[1] 25 frames per second using 352 × 288 CIF images compressed using MPEG4
(http://www.info4security.com/story.asp?storyCode=3093501)

D. Duke et al. (Eds.): SAMT 2008, LNCS 5392, pp. 105–120, 2008.
© Springer-Verlag Berlin Heidelberg 2008

includes a mapping between behaviours and the semantic concepts which encapsulate them. In essence, automated techniques suffer from issues presented by the multimedia semantic gap[44], between semantic queries which users readily express and which systems cannot answer.

Although some efforts have attempted to bridge this gap for behavioural descriptions, an area which has received little attention is semantic appearance descriptions, especially in surveillance. Semantic whole body descriptions (Height, Figure etc.) and global descriptions (Sex, Ethnicity, Age, etc.) are a natural way to describe individuals. Their use is abundant in character description in narrative, helping readers put characters in a richer context with a few key words such as *slender* or *stout*. In a more practical capacity, stable physical descriptions are of key importance in eyewitness crime reports, a scenario where human descriptions are paramount as high detail images of assailants are not always available. Many important semantic features are readily discernible from surveillance videos by humans, and yet are challenging to extract and analyse automatically. Unfortunately, the manual annotation of videos is a laborious[7][16] process, too slow for effective use in real time CCTV footage and vulnerable to various sources of human error (subject variables, anchoring etc.). Automatic analysis of the way people walk[29] (their gait) is an efficient and effective approach to describing human features at a distance. Yet automatic gait analysis techniques do not necessarily generate signatures which are immediately comprehensible by humans. We show that Latent Semantic Analysis techniques, as used successfully by the image retrieval community, can be used to associate semantic physical descriptions with automatically extracted gait features. In doing so, we contend that retrieval tasks involving semantic physical descriptions could be readily facilitated.

The rest of this paper is organised in the following way. In Section 2 we describe Latent Semantic Analysis, the technique chosen to bridge the gap between semantic physical descriptions and gait signatures. In Section 3 we introduce the semantic physical *traits* and their associated *terms*; justifying their psychological validity. In Section 4 we briefly summarise modern gait analysis techniques and the gait signature chosen for our experiments. In Section 5 we outline the source of our experiment's description data, using it in Section 6 where we outline the testing methodology and show that our novel approach allows for content-based video retrieval based on gait. Finally in Section 7 we discuss the final results and future work.

2   Latent Semantic Analysis

2.1   The Singular Value Decomposition

In text retrieval, Cross Language Latent Semantic indexing (CL-LSI) [20], itself an extension of LSI [9], is a technique which statistically relates contextual-usage of terms in large corpuses of text documents. In our approach, LSI is used to construct a Linear-Algebraic Semantic Space from multimedia sources[14][37] within which documents and terms sharing similar *meaning* also have similar *spacial location*.

We start by constructing an occurrence matrix $\mathbf{O}$ whose values represent the *presence* of terms in documents (columns represent documents and rows represent terms). In our scenario documents are videos. Semantic features and automatic features are considered terms. The "occurrence" of an automatic feature signifies the magnitude of that portion of the automatic feature vector while the "occurrence" of a semantic term signifies its semantic relevance to the subject in the video. Our goal is the production of a rank reduced factorisation of the observation matrix consisting of a term matrix $\mathbf{T}$ and document matrix $\mathbf{D}$, such that:

$$\mathbf{O} \approx \mathbf{TD}. \tag{1}$$

Where the vectors in $\mathbf{T}$ and $\mathbf{D}$ represent the *location* of individual terms and documents respectively within some shared space.

$\mathbf{T}$ and $\mathbf{D}$ can be efficiently calculated using the singular value decomposition (SVD) which is defined as:

$$\mathbf{O} = \mathbf{U\Sigma V}^T \tag{2}$$

Such that $\mathbf{T} = \mathbf{U}$ and $\mathbf{D} = \mathbf{\Sigma V}^T$, and the rows of $\mathbf{U}$ represent positions of terms and the columns of $\mathbf{\Sigma V}^T$ represent the position of documents. The diagonal entries of $\mathbf{\Sigma}$ are equal to the singular values of $\mathbf{O}$. The columns of $\mathbf{U}$ and $\mathbf{V}$ are, respectively, left- and right-singular vectors for the corresponding singular values in $\mathbf{\Sigma}$. The singular values of any $m \times n$ matrix $\mathbf{O}$ are defined as values $\{\sigma_1, .., \sigma_r\}$ such that :

$$\mathbf{O}\mathbf{v_i} = \sigma_i \mathbf{u_i}, \tag{3}$$

and

$$\mathbf{O}^T \mathbf{u_i} = \sigma_i \mathbf{v_i} \tag{4}$$

Where $\mathbf{v}_i$ and $\mathbf{u}_i$ are defined as the right and left singular vectors respectively.

In can be shown that $\mathbf{v}_i$ and $\mathbf{u}_i$ are in fact the *eigenvectors* with corresponding *eigenvalues* $\{\lambda_1 = \sigma_1^2, .., \lambda_r = \sigma_r^2\}$ of the square symmetric matrices $\mathbf{O}^T\mathbf{O}$ and $\mathbf{O}\mathbf{O}^T$ respectively, referred to as the *co-occurrence* matrices. The matrix $\mathbf{U}$ contains all the eigenvectors of $\mathbf{O}\mathbf{O}^T$ as its rows while $\mathbf{V}$ contains all the eigenvectors of $\mathbf{O}^T\mathbf{O}$ its rows and $\mathbf{\Sigma}$ contains all the eigenvalues along its diagonal. Subsequently:

$$\mathbf{O}^T\mathbf{O} = \mathbf{V\Sigma}^T\mathbf{U}^T\mathbf{U\Sigma V}^T = \mathbf{V\Sigma}^T\mathbf{\Sigma V}^T, \tag{5}$$

$$\mathbf{O}\mathbf{O}^T = \mathbf{U\Sigma V}^T\mathbf{V\Sigma}^T\mathbf{U}^T = \mathbf{U\Sigma\Sigma}^T\mathbf{U}^T. \tag{6}$$

To appreciate the importance of SVD and the eigenvector matrices $\mathbf{V}$ and $\mathbf{U}$ for information retrieval purposes, consider the *meaning* of the respective co-occurrence matrices.

$$\mathbf{T_{co}} = \mathbf{O}\mathbf{O}^T, \tag{7}$$

$$\mathbf{D_{co}} = \mathbf{O}^T\mathbf{O}. \tag{8}$$

The magnitude of the values in $\mathbf{T_{co}}$ relate to how often a particular term appears with every other term throughout all documents, therefore some concept of the "relatedness" of terms. The values in $\mathbf{D_{co}}$ relate to how many terms every document shares with every other document, therefore the "relatedness" of documents.

By definition the matrix of eigenvectors $\mathbf{U}$ and $\mathbf{V}$ of the two matrices $\mathbf{T_{co}}$ and $\mathbf{D_{co}}$ respectively form two basis for the co-occurrence spaces, i.e. the combination of terms (or documents) which the entire space of term co-occurrence can be projected into without information loss.

In a similar strategy to Principal Components Analysis (PCA), LSA works on the premise that the eigenvectors represent underlying latent concepts encoded by the co-occurrence matrix and by extension the original data. It is helpful to think of these latent concepts as mixtures (or weightings) of terms or documents. Making such an assumption allows for some interesting mathematical conclusions. Firstly, the eigenvectors with the largest corresponding eigenvalues can be thought of the *most* representative latent concepts of the space. This means by using only the most relevant components of $\mathbf{T}$ and $\mathbf{D}$ (as ordered by the singular values), less meaningful underlying concepts can be ignored and higher accuracy achieved. Also as both the document and term co-occurrence matrices represent the same data, their latent concepts must be identical and subsequently comparable[2]. Therefore the position of every term or document projected into the latent space are similar if the terms and documents in fact share similar meaning.

2.2   Using SVD

With this insight, our tasks becomes the choice of semantic and visual terms to be observed from each subject for the generation of an observation matrix. Once this matrix is generated, content-based retrieval by semantic query of unannotated documents can be achieved by exploiting the projection of partially observed vectors into the eigenspace represented by either $\mathbf{T}$ or $\mathbf{D}$.

Assume we have two subject-video collections, a fully annotated training collection and a test collection, lacking semantic annotations. A matrix $\mathbf{O}_{train}$ is constructed such that training documents are held in its columns. Both visual and semantic terms are fully observed for each training document, i.e. a term is set to a non-zero value encoding its existence or relevance to a particular video. Using the process described in Section 2.1 we can obtain $\mathbf{T}_{train}$ and $\mathbf{D}_{train}$ for the training matrix $\mathbf{O}_{train}$.

Content-Based Retrieval. To retrieve the set of unannotated subjects based on their visual gait components alone, a new partially observed document matrix $\mathbf{O}_{test}$ is constructed such that visual gait terms are prescribed and semantic terms are set to zero. For retrieval by semantic terms, a query document matrix is constructed where all visual and non-relevant semantic terms are set to zero

---

[2] It can also be shown that the two sets of eigenvectors are in fact in the same vector space[37] and are subsequently directly comparable.

while relevant semantic terms are given a non-zero value (usually 1.0), this query matrix is $\mathbf{O}_{query}$. The query and test matrix are projected in the latent space in following manner:

$$\mathbf{D}_{test} = \mathbf{T}_{train}^T \mathbf{O}_{test}, \qquad (9)$$
$$\mathbf{D}_{query} = \mathbf{T}_{train}^T \mathbf{O}_{query}. \qquad (10)$$

Projected test documents held in $\mathbf{D}_{test}$ are simply ordered according to their co-sine distance from query documents in $\mathbf{D}_{query}$ for retrieval. This process readily allows for automatic annotation, though exploration in this area is beyond the scope of this report. We postulate that annotation could be achieved by finding the distance of $\mathbf{D}_{test}$ to each term in $\mathbf{T}_{train}$. A document is annotated with a term if that term is the closest compared to others belonging to the same physical *trait* (discussed in more detail in Section 3).

We show results for retrieval experiments in Section 6.

3   Human Physical Descriptions

The description of humans based on their physical features has been explored for several purposes including medicine[34], eyewitness analysis and human iden-tification [3]. Descriptions chosen differ in levels of granularity and include fea-tures both visibly measurable but also those only measurable through use of specialised tools. One of the first attempts to systematically describe people for identification based on their physical traits was the anthropometric system developed by Bertillon [5] in 1896. His system used eleven precisely measured traits of the human body including height, length of right ear and width of cheeks. This system was quickly surpassed by other forms of forensic analysis such as fingerprints. More recently, physical descriptions have also been used in biometric techniques as an ancillary data source where they are referred to as *soft biometrics*[28], as opposed to primary biometric sources such as iris, face or gait. In behaviour analysis, several model based techniques[1] attempt the automatic extraction of individual body components as a source of behavioural information. Though the information about the individual components is not used directly, these techniques provide some insight into the level of granularity at which body features are still discernible at a distance.

When choosing the features that should be considered for semantic retrieval of surveillance media, two major questions must be answered. Firstly, which human traits should be described and secondly, how should these traits be rep-resented. The following sections outline and justify the traits chosen and outline the semantic terms chosen for each physical trait.

Physical Traits

To match the advantages of automatic surveillance media, one of our primary concerns was to choose traits that are discernible by humans at a distance. To

---
[3] Interpol. Disaster Victim Identification Form (Yellow). booklet, 2008.

do so we must firstly ask which traits individuals can *consistently* and *accurately* notice in each other at a distance. Three independent traits - Age, Race and Sex, are agreed to be of primary significance in cognitive psychology. For gait, humans have been shown to successfully perceive such categories using generated point light experiments [39] with limited visual cues. Other factors such as the target's perceived somatotype [26] (build or physique attributes) are also prominent in cognition.

In the eyewitness testimony research community there is a relatively mature idea of which concepts witnesses are most likely to recall when describing individuals [42]. Koppen and Lochun [19] provide an investigation into witness descriptions in archival crime reports. Not surprisingly, the most accurate and highly mentioned traits were Sex (95% mention 100% accuracy), Height (70% mention 52% accuracy), Race (64% mention 60% accuracy) and Skin Colour (56% mention, accuracy not discussed). Detailed head and face traits such as Eye Shape and Nose Shape are not mentioned as often and when they are mentioned, they appear to be inaccurate. More prominent head traits such as Hair Colour and Length are mentioned more consistently, a result also noted by Yarmey and Yarmey [43]. Descriptive features which are visually prominent yet less permanent (e.g. clothing) often vary with time and are of less interest than other more permanent physical traits.

Traits regarding build are of particular interest, having a clear relationship with gait while still being reliably recalled by eyewitnesses at a distance. Few studies thus far have attempted to explore build in any amount of detail beyond the brief mention of Height and Weight. MacLeod et al. [25] performed a unique analysis on whole body descriptions using bipolar scales to define traits. Initially, whole body traits often described by people in freeform annotations experiments were gauged using a set of moving and stationary subjects. From an initial list of 1238 descriptors, 23 were identified as unique and formulated as five-point bipolar scales. The reliability and descriptive capability of these features were gauged in a separate experiment involving subjects walking at a regular pace around a room. Annotations made using these 23 features were assessed using product moment correlation and their underlying similarity was assessed using a principal components analysis. The 13 most reliable terms and most representative of the principle components have been incorporated into our final set of traits.

Jain et al. [17] outline a set of key characteristics which determine a physical trait's suitability for use in biometric identification, a comparable task to multimedia retrieval. These include: Universality, Distinctiveness, Permanence and Collectability.

The choice of our physiological traits keeps these tenets in mind. Our semantic descriptions are universal in that we have chosen factors which everyone has. We have selected a set of subjects who appeared to be semantically distinct in order to confirm that these semantic attributes can be used. The descriptions are relatively permanent: overall Skin Colour naturally changes with tanning, but our description of Skin Colour has racial overtones and these are perceived to be more constant. Our attributes are easily collectible and have been specifically selected for being easily discernible at a distance by humans. However much care has been taken over procedure and definition to ensure consistency of acquisition (see Section 5).

Using a combination of the studies in cognitive science, witness descriptions and the work by MacLeod et al. [25] we generated a list of visual semantic traits which is given in Table 1.

Semantic Terms

Having outlined which physical traits should allowed for, the next question is how these traits should be represented. Soft biometric techniques use a mixture of categorical metrics (e.g. Ethnicity) and value metrics (e.g. Height) to represent their traits. Humans are generally less consistent when making value judgements in comparison to category judgements. Subsequently, in our approach we formulate all traits with sets of mutually exclusive semantic terms rather than using value metrics. This approach is more representative of the categorical nature of human cognition [38] [26] [39]. This is naturally achieved for certain traits, primarily when no applicable underlying value order exists (Sex, Hair Colour etc.). For other traits representable with intuitive value metrics (Age, Lengths, Sizes etc.) bipolar scales representing concepts from *Small* to *Large* are used as semantic terms. This approach closely matches human categorical perception. Annotations obtained from such approaches have been shown to correlate with measured numerical values [8]. Perhaps the most difficult trait for which to find a limited set of terms was Ethnicity. There is a large corpus of work [12] [33] [2] exploring ethnic classification, each outlining different ethnic terms; ranging from the use of 3 to 200, with non necessarily convergent. Our ethnic terms encompass the three categories mentioned most often and an extra two categories (Indian and Middle Eastern) matching the UK census[4].

4   Automatic Gait Descriptions

In the medical, psychological and biometric community, automatic gait recognition has enjoyed considerable attention in recent years. Psychological significance in human identification has been demonstrated by various experiments [39] [18]; it is clear that the way a person walks and their overall structure hold a significant amount of information used by humans when identifying each other. Inherently, gait recognition has several attractive advantages as a biometric. It is unobtrusive, meaning people are more likely to accept gait analysis over other, more accurate, yet more invasive biometrics such as finger print recognition or iris scans. Also gait is one of the few biometrics which has been shown to identify individuals effectively at large distances and low resolutions. However this flexibility also gives rise to various challenges in the use of gait as a biometric. Gait is (in part) a behavioural biometric and as such is affected by a large variety of

---

[4] http://www.statistics.gov.uk/about/Classifications/
ns_ethnic_classification.asp Ethnic classification

Table 1. Physical traits and associated semantic terms

Body Shape		Global	
1. Arm Length	[Very Short, Short, Average, Long, Very Long]	14. Age	[Infant, Pre Adolescence, Adolescence, Young Adult, Adult, Middle Aged, Senior]
2. Arm Thickness	[Very Thin, Thin, Average, Thick, Very Thick]	15. Ethnicity	[Other, European, Middle Eastern, Far Eastern, Black, Mixed]
3. Chest	[Very Slim, Slim, Average, Large, Very Large]	16. Sex	[Female, Male]
4. Figure	[Very Small, Small, Average, Large, Very Large]	17. Skin Colour	[White, Tanned, Oriental, Black]
5. Height	[Very Short, Short, Average, Tall, Very Tall]	**Head**	
		18. Facial Hair Colour	[None, Black, Brown, Blond, Red, Grey]
6. Hips	[Very Narrow, Narrow, Average, Broad, Very Broad]	19. Facial Hair Length	[None, Stubble, Moustache, Goatee, Full Beard]
7. Leg Length	[Very Short, Short, Average, Long, Very Long]	20. Hair Colour	[Black, Brown, Blond, Grey, Red, Dyed]
8. Leg Shape	[Very Straight, Straight, Average, Bow, Very Bowed]	21. Hair Length	[None, Shaven, Short, Medium, Long]
9. Leg Thickness	[Very Thin, Thin, Average, Thick, Very Thick]	22. Neck Length	[Very Short, Short, Average, Long, Very Long]
10. Muscle Build	[Very Lean, Lean, Average, Muscly, Very Muscly]	23. Neck Thickness	[Very Thin,Thin,Average,Thick,Very Thick]
11. Proportions	[Average, Unusual]		
12. Shoulder Shape	[Very Square, Square, Average, Rounded, Very Rounded]		
13. Weight	[Very Thin, Thin, Average, Fat, Very Fat]		

co-variates including mood, fatigue, clothing etc. all of which can result in large within-subject (intra-class) variance.

Over the past 20 years there has been a considerable amount of work dedicated to effective automatic analysis of gait with the use of marker-less machine vision techniques attempting to match the capabilities of human gait perception[30]. Broadly speaking, these techniques can be separated into model based techniques and holistic statistical techniques.

The latter approaches tend to analyse the human silhouette and its temporal variation without making any assumptions as to how humans tend to move. An early example of such an approach was performed by Little and Boyd [23] who extract optic flow "blobs" between frames of a gait video which they use to fit an ellipsoids to describe predominant axis of motion. Murase and Sakai [27] analyse gait videos by projecting each frame's silhouettes into the eigenspace separately and using the trajectory formed by all of an individual's separate frames in the eigenspace as their signature. Combining each frame silhouette and averaging by number of frames, or simply average silhouette [13] [24] [40], is the most popular holistic approach. It provides relatively promising results and is comparatively simple to implement and as such is often used as a baseline algorithm.

Model based techniques start with some assumption of how humans move or a model for human body structure, usually restricted to one view point, though some tackle the problem in 3D. Values for model parameters are estimated which most faithfully represent the sensed video data. An elegant early approach by [31] stacked individual silhouettes in an x-y-time (XYT) space, fitting a helix to the distinctive pattern caused by human legs at individual XT slices. The helix perimeters are used to define the parameters for a five-part stick model. Another, more recent approach by BenAbdelkader et al. [3] uses a structural model and attempts to gather evidence for subject height and cadence.

Model based techniques make several assumptions and explicitly extract certain information from subject videos. Though this would be useful for specific structural semantic terms (Height, Arm/Leg dimensions etc.), the model could feasibly ignore global semantic terms (Sex, Ethnicity etc.) evidence for which could exist in the holistic information[21]. Subsequently we choose the simple yet powerful average silhouette operation for our automatic gait signature both for purposes of simplicity and to increase the likelihood of correlation with global semantic terms.

5   Semantic and Automatic Data Source

In this section we describe the procedures undertaken to extract automatic and manual data sources describing our gait videos. Our videos are of 115 individual subjects each with a minimum of 6 video samples from the Southampton University Gait Database [36] [36]. In our experiments, the videos used are from camera set-up "a" during which subjects walk at a natural pace side on to the plane of the camera view and walking either towards the left or right. Each subject has been annotated by at least two separate annotators, though 10 have been annotated with 40 annotators as part of a previous, more rigourous, though smaller scale experiment [35].

Semantic Features

Semantic annotations were collected using the GaitAnnotate system; a web based application designed to show arbitrary biometric data sources to users for annotation, as shown in Fig. 1. This interface allows annotators to view all video samples of a subject as many times as they require. Annotators were asked to describe subjects by selecting semantic terms for each physical trait. They were instructed to label *every* trait for *every* subject and that each trait should be completed with the annotator's own notions of what the trait *meant*. Guidelines were provided to avoid common confusions e.g. that Height of an individual should be assigned absolutely in compared to a perceived global "Average" where traits such as Arm Length could be annotated in comparison to the subject's overall physique. This annotation data was also gathered from some subjects present in the video set, as well as from subjects not present (e.g. a class of Psychology students, the main author etc.).

To gauge an upper limit for the quality of semantic retrieval, we strive to assure the semantic data is of optimal quality. The annotation gathering process was designed to carefully avoid (or allow the future study of) inherent weaknesses and inaccuracies present in human generated descriptions. The error factors that the system accommodates include:

- **Memory[10]** - Passage of time may affect a witness' recall of a subject's traits. Memory is affected by variety of factors e.g. the construction and utterance of featural descriptions rather than more accurate (but indescribable) holistic descriptions. Such attempts often alter memory to match the featural descriptions.
- **Defaulting[22]** - Features may be left out of descriptions in free recall. This is often not because the witness failed to remember the feature, but rather that the feature has some default value. Race may be omitted if the crime occurs in a racially homogenous area, Sex may be omitted if suspects are traditionally Male.
- **Observer Variables[11][32]** - A person's own physical features, namely their self perception and mental state, may affect recall of physical variables. For example, tall people have a skewed ability to recognise other tall people but will have less ability when it comes to the description shorter individuals, not knowing whether they are average or very short.
- **Anchoring[6]** - When a person is asked a question and is initially presented with some default value or even seemingly unrelated information, the replies given are often weighted around those initial values. This is especially likely when people are asked for answers which have some natural ordering (e.g. measures of magnitude)

We have designed our semantic data gathering procedure to account for all these factors. Memory issues are addressed by allowing annotators to view videos of subjects as many times as they please, also allowing them to repeat a particular video if necessary. Defaulting is avoided by explicitly asking individuals for each trait outlined in Table 1, this means that even values for apparently *obvious* traits are filled in and captured. This style of interrogative description, where constrained responses are explicitly requested, is more complete than free-form narrative recall but may suffer from inaccuracy, though not to a significant degree [43]. Subject variables can never be completely removed so instead we allow the study of differing physical traits across various annotators. Users are asked to self annotate based on self perception, also certain subjects being annotated are themselves annotators. This allows for some concept of the annotator's own appearance to be taken into consideration when studying their descriptions of other subjects. Anchoring can occur at various points of the data capture process. We have accounted for anchoring of terms gathered for individual traits by setting the default term of a trait to a neutral "Unsure" rather than any concept of "Average".

To allow for inclusion of semantic terms of each trait in the LSA observation matrix, each semantic term is represented by its occurrence for each subject. This occurrence is extracted by finding a consensus between annotators which

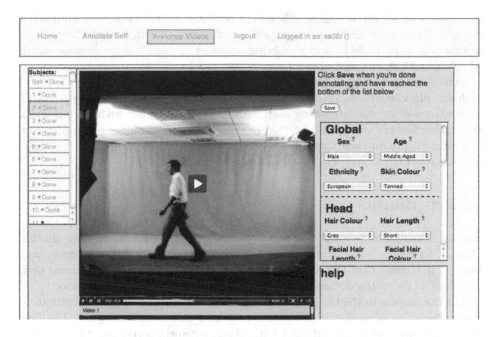

Fig. 1. Example of GAnn interface

made a judgement of a particular term for a particular subject. Each of the n annotators produces the the ith annotation assigning the jth term for the kth subject, producing a response $r_{ijk} \in [0, 1]$. The value for each term t_{jk} for j^{th} for the kth subject is calculated such that:

$$t_{jk} = \frac{1}{n} \sum_{i=0}^{n} r_{ij} \qquad (11)$$

This results in a single annotation for each subject for each term which is a value between 0.0 and 1.0 which defines how relevant a particular semantic term is to a particular subject, i.e. its occurrence (see Section 2).

If an annotator responds as with "Unsure" for each trait, or does not provide the annotations at all, their response is set to the mode of that trait across all annotators across that particular subject. This results in a complete 113x115 (113 semantic terms, 115 subjects) matrix which is concatenated with the automatic feature matrix described in the following section.

Automatic Gait Features

The automatic feature vector used for these experiments were the average silhouette gait signatures. For each gait video, firstly the subject is extracted from the scene with a median background subtraction and transformed into a binary silhouette. This binary silhouette is resized to a 64x64 image to make the signature distance invariant. The gait signature of a particular video is the averaged summation of all these binary silhouettes across one gait cycle. For simplicity the gait signature's intensity values are use directly, although there have been several attempts made to find significant features in such feature vectors, using ANOVA or PCA [40] and also a Fourier Decomposition [15].

This results in 4096 (64x64) automatic feature components which describe each sample video of each of the 115 subjects. The final observation matrix $\mathbf{O}$ is constructed by concatenating each sample feature vector with its subject's annotation feature vector as described in the previous section. This complete set of automatically and semantically observed subjects is manipulated in Section 6 to generate $\mathbf{O}_{train}$ and $\mathbf{O}_{test}$ as described in Section 2.

6   Experiments

For the retrieval experiment it was required to construct a training matrix $\mathbf{O}_{train}$, for which visual features and semantic features are fully observed, and $\mathbf{O}_{test}$ matrix such that the semantic features are set to zero. The retrieval task attempts to order the documents in $\mathbf{O}_{test}$ against some semantic queries.

The documents in the training stage are the samples (and associated semantic annotations) of a randomly selected set of half of the 115 subjects, the test documents are the other subjects with their semantic terms set to zero. For analysis, 10 such sets are generated and latent semantic spaces ($\mathbf{T}_{train}$ and $\mathbf{D}_{train}$) are generated for each.

6.1   Semantic Query Retrieval Results

We test the retrieval ability of our approach by testing each semantic term in isolation (e.g. Sex Male, Height Tall etc.). A few example retrieval queries can be seen in Fig. 3. Here, the Male subjects have been retrieved successfully as have the Female subjects. In Pre-Adolescence, the system selects two children but one adult, incorrectly. The Hair Length retrieval is consistently correct. To put our results in context we also measure the standard mean average precision (mAP) metric as calculated by TREC-Eval. The mAP of each semantic term is taken from the mAP of a random ordering for each query. To generate the random mAP we generate 100 random orderings for each semantic query and average their mAP.Fig. 2 shows the sum of the random order difference of each semantic terms for each trait. These results give some idea of which traits our approach is most capable of performing queries against, and which it is not.

Our results show some merit and produce both success and failure, as expected. It has been shown in previous work for example that Sex (mAP=0.12) is decipherable from average silhouettes alone [21], achieved by analysing the separate parts of the human silhouette. It is also expected that physical metrics of mass such as Weight (mAP=0.043), Figure (mAP=0.041) and Leg Thickness (mAP=0.044) were also likely to be relatively successful as the average silhouette maintains a linear representation of these values in the overall intensity of pixels. Also, the poor performance of Height (mAP=0.0026) is expected as the

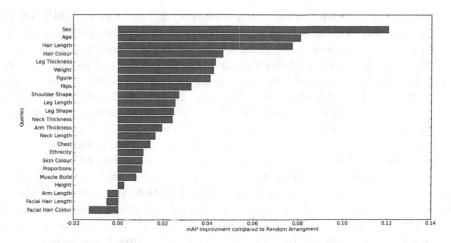

Fig. 2. Bar-graph showing the mean average precision improvement for each semantic trait. Each trait is the weighted sum of its substituent semantic terms.

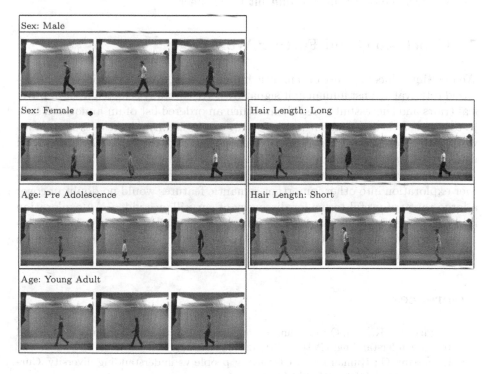

Fig. 3. Some Retrieval Results

average silhouette features have removed this feature by normalising silhouettes, making them the same height for comparison. Only a latent concept of Height in terms of the aspect ratio is maintained.

Perhaps most surprising is the relative success of Hair Colour (mAP=0.047) metric as, on first inspection it seems as though silhouette images maintain no colour information. However, the construction of binary silhouettes is undoubtedly affected by hair colour when compared to background, and as such the average silhouette images retain hair colour as brightness in the head region. Veres et al. [40] noted that this region was the most useful portion of the silhouette for recognition. It is also likely that this Hair Colour holds some significant relationship with another semantic feature, for example Sex, as most of our Female participants were indeed of East Asian origin and subsequently had Black Hair. In future work we discuss an exploration into other such points of interest.

A t-test was also performed to gauge the significance of each mAP difference from random in relation to the mAP standard deviation. A small p-value indicates higher significance. The p-value of each mAP difference shows that many of our retrieval precision rates were significant[5]. Sex, Hair Colour/Length and Age each shared a p-value of 10^{-5} where Facial Hair features had p-values > 0.2. This further demonstrates the merit of our approach. It should be noted that there is a correlation with traits that performed poorly and those reported by annotators to be confusing and difficult to decipher.

7   Conclusions and Further Work

We have introduced the use of semantic human descriptions as queries in content-based retrieval against human gait signatures. We carefully selected a set of physical traits and successfully used them return an ordered list of un-annotated subjects based on their gait signature alone. Our results confirm the results of previous work with regards to traits such as Sex and we also note the capability of retrieval using other traits, previously unexplored, such as Age and some build attributes.

There are several interesting avenues of research suggested by this work. A further exploration into other important semantic features would no doubt uncover a large range of useful terms for discovery of surveillance video. An exploration into other gait signatures would also improve the recall of certain semantic features. Using model based techniques to more directly extract Height and limb attributes would no doubt improve their retrieval rates.

References

[1] Aggarwal, J.K., Cai, Q.: Human motion analysis: A review. Computer Vision and Image Understanding: CVIU 73(3), 428–440 (1999)
[2] Barbujani, G.: Human races: Classifying people vs understanding diversity. Current Genomics 6(12), 215–226 (2005)
[3] BenAbdelkader, C., Cutler, R., Davis, L.: Stride and cadence as a biometric in automatic person identification and verification. In: Proc. 5th IEEEFG, pp. 372–377 (May 2002)

---

[5] To a threshold of 0.01.

[4] Bennetto, J.: Big brother britain 2006: we are waking up to a surveillance society all around us. The Independant (2006)
[5] Bertillon, A.: Signaletic Instructions including the theory and practice of Anthropometrical Identification. The Werner Company (1896)
[6] Chapman, G.B., Johnson, E.J.: Incorporating the irrelevant: Anchors in judgments of belief and value. In: Heuristics and Biases: The Psychology of Intuitive Judgment, pp. 120–138. Cambridge University Press, Cambridge (2002)
[7] Davies, A., Velastin, S.: A Progress Review of Intelligent CCTV Surveillance Systems. In: IEEEIDAACS 2005, pp. 417–423 (September 2005)
[8] Dawes, R.M.: Suppose We Measured Height With Rating Scales Instead of Rulers. App. Psych. Meas. 1(2), 267–273 (1977)
[9] Deerwester, S.C., Dumais, S.T., Landauer, T.K., Furnas, G.W., Harshman, R.A.: Indexing by latent semantic analysis. J. of the American Society of Information Science 41(6), 391–407 (1990)
[10] Ellis, H.D.: Practical aspects of facial memory. In: Eyewitness Testimony: Psychological perspectives, section 2, pp. 12–37. Cambridge University Press, Cambridge (1984)
[11] Flin, R.H., Shepherd, J.W.: Tall stories: Eyewitnesses ability to estimate height and weight characteristics. Human Learning 5 (1986)
[12] Gould, S.J.: The Geometer of Race. Discover, pp. 65–69 (1994)
[13] Han, J., Bhanu, B.: Statistical feature fusion for gait-based human recognition. In: Proc. IEEE CVPR 2004, vol. 2 , pp. II–842–II–847 (June-July 2004)
[14] Hare, J.S., Lewis, P.H., Enser, P.G.B., Sandom, C.J.: A linear-algebraic technique with an application in semantic image retrieval. In: Sundaram, H., Naphade, M., Smith, J.R., Rui, Y. (eds.) CIVR 2006. LNCS, vol. 4071, pp. 31–40. Springer, Heidelberg (2006)
[15] Hayfron-Acquah, J.B., Nixon, M.S., Carter, J.N.: Automatic Gait Recognition by Symmetry Analysis. Pattern Recognition Letters 24(13), 2175–2183 (2003)
[16] Hu, W., Tan, T., Wang, L., Maybank, S.: A survey on visual surveillance of object motion and behaviors. IEEETSMC(A) 34(3), 334–352 (2004)
[17] Jain, A.K., Ross, A., Prabhakar, S.: An Introduction to Biometric Recognition. Trans. CSVT 14, 4–19 (2004)
[18] Johansson, G.: Visual perception of biological motion and a model for its analysis. Percept. Phychophys. 14(2), 201–211 (1973)
[19] Koppen, P.V., Lochun, S.K.: Portraying perpetrators; the validity of offender descriptions by witnesses. Law and Human Behavior 21(6), 662–685 (1997)
[20] Landauer, T., Littman, M.: Fully automatic cross-language document retrieval using latent semantic indexing. In: 6th Annual Conference of the UW Centre for the New OED, pp. 31–38 (1990)
[21] Li, X., Maybank, S., Yan, S., Tao, D., Xu, D.: Gait components and their application to gender recognition. IEEETSMC(C) 38(2), 145–155 (2008)
[22] Lindsay, R., Martin, R., Webber, L.: Default values in eyewitness descriptions. Law and Human Behavior 18(5), 527–541 (1994)
[23] Little, J., Boyd, J.: Describing motion for recognition. In: SCV 1995, page 5A Motion II (1995)
[24] Liu, Z., Sarkar, S.: Simplest representation yet for gait recognition: averaged silhouette. In: Proc. ICPR 2004, vol. 4, pp. 211–214 (August 2004)
[25] MacLeod, M.D., Frowley, J.N., Shepherd, J.W.: Whole body information: Its relevance to eyewitnesses. In: Adult Eyewitness Testimony, ch. 6. Cambridge University Press, Cambridge (1994)

[26] Macrae, C.N., Bodenhausen, G.V.: Social Cognition: Thinking Categorically about Others. Ann. Review of Psych. 51(1), 93–120 (2000)

[27] Murase, H., Sakai, R.: Moving object recognition in eigenspace representation: gait analysis and lip reading. Pattern Recogn. Lett. 17(2), 155–162 (1996)

[28] Nandakumar, K., Dass, S.C., Jain, A.K.: Soft biometric traits for personal recognition systems. In: Zhang, D., Jain, A.K. (eds.) ICBA 2004, vol. 3072, pp. 731–738. Springer, Heidelberg (2004)

[29] Nixon, M., Carter, J.N.: Automatic recognition by gait. Proc. of the IEEE 94(11), 2013–2024 (2006)

[30] Nixon, M.S., Carter, J.N.: Automatic recognition by gait. Proceedings of the IEEE 94(11), 2013–2024 (2006)

[31] Niyogi, S., Adelson, E.: Analyzing and recognizing walking figures in XYT. In: Proc. CVPR 1994, pp. 469–474 (June 1994)

[32] O'Toole, A.J.: Psychological and Neural Perspectives on Human Face Recognition. In: Handbook of Face Recognition. Springer, Heidelberg (2004)

[33] Ponterotto, J.G., Mallinckrodt, B.: Introduction to the special section on racial and ethnic identity in counseling psychology: Conceptual and methodological challenges and proposed solutions. J. of Counselling Psych. 54(3), 219–223 (2007)

[34] Rosse, C., Mejino, J.L.V.: A reference ontology for biomedical informatics: the foundational model of anatomy. J. of Biomed. Informatics 36(6), 478–500 (2003)

[35] Samangooei, S., Guo, B., Nixon, M.S.: The use of semantic human description as a soft biometric. In: BTAS (September 2008)

[36] Shutler, J., Grant, M., Nixon, M.S., Carter, J.N.: On a large sequence-based human gait database. In: RASC 2006, pp. 66–72 (2002)

[37] Skillicorn, D.: Understanding Complex Datasets. In: Singular Value Decomposition (SVD), ch. 3. Chapman & Hall/CRC (2007)

[38] Tajfel, H.: Social Psychology of Intergroup Relations. Ann. Rev. of Psych. 33, 1–39 (1982)

[39] Troje, N.F., Sadr, J., Nakayama, K.: Axes vs averages: High-level representations of dynamic point-light forms. Vis. Cog. 14, 119–122 (2006)

[40] Veres, G., Gordon, L., Carter, J., Nixon, M.: What image information is important in silhouette-based gait recognition? In: Proc. IEEE CVPR 2004, vol. 2, pp. II-776–II-782 (June-July 2004)

[41] Vrusias, B., Makris, D., Renno, J.-P., Newbold, N., Ahmad, K., Jones, G.: A framework for ontology enriched semantic annotation of cctv video. In: WIAMIS 2007, p. 5 (June 2007)

[42] Wells, G.L., Olson, E.A.: Eyewitness testimony. Ann. Rev. of Psych. 54, 277–295 (2003)

[43] Yarmey, A.D., Yarmey, M.J.: Eyewitness recall and duration estimates in field settings. J. of App. Soc. Psych. 27(4), 330–344 (1997)

[44] Zhao, R., Grosky, W.: Bridging the Semantic Gap in Image Retrieval. IEEE Transactions on Multimedia 4, 189–200 (2002)

Context as a Non-ontological Determinant of Semantics

Simone Santini and Alexandra Dumitrescu

Escuela Politécnica Superior, Universidad Autónoma de Madrid

Abstract. This paper proposes an alternative to formal annotation for the representation of semantics. Drawing on the position of most of last century's linguistics and interpretation theory, the article argues that meaning is not a property of a document, but an outcome of a contextualized and situated process of interpretation. The consequence of this position is that one should not quite try to represent the meaning of a document (the way formal annotation does), but the context of the activity of which search is part.

We present some general considerations on the representation and use of the context, and a simple example of a technique to encode the context represented by the documents collected in the computer in which one is working, and to use them to direct search. We show preliminary results showing that even this rather simpleminded context representation can lead to considerable improvements with respect to commercial search engines.

1   Introduction

To be critical of ontology today is only marginally less dangerous than being a royalist in Paris in 1793. One doesn't quite risk one's neck, but has the feeling that his professional reputation might end up summarily guillotined. The issue is particularly poignant here, because this paper will be, essentially, an argument against the use of formal annotation (the framework in which ontology is used) and against the current orientation of the semantic web. Formal annotation is based on the idea that meaning is somehow contained in a document. This position doesn't take into account that whatever semantic there might be in a data access (and it is probably not that much) comes from the interaction of a reader with the document, an interaction that takes place in a context and using rules determined by the activity of which it is part. Consequently, *querying* can be seen as an isolated activity, independent of the substratum on which it takes place, very much the way it is in structured data bases. But whoever makes enthusiastic claims about the new "richness" of the data that can be found on the Internet should be coherent and admit that the very nature of these data will make old concepts (including the concepts of query and search) obsolete.

We will present an alternative model based on two assumptions: firstly, that the meaning of a document is given by the change it provokes in the context of the activity in which it is read; secondly, that these activities can be configured as *games*, and that what is usually called a query is but a type of move in these games.

Some people might find the arguments a bit on the philosophical side, but it must be remembered that semantics is essentially a philosophical issue. The computer scientist who designs programs for particle accelerators can't help but coming in touch with a

D. Duke et al. (Eds.): SAMT 2008, LNCS 5392, pp. 121–136, 2008.

bit of quantum physics, because that is what particle physics is about. Analogously, computing scientists who want to work on semantics can't help but deal with (as Sam Spade would have put it) the stuff semantics is made of: philosophy.

2   Ontology for the Representation of Semantics

One solution to the problem of semantic data processing, quite fashionable in the computing milieu these days, entails the *formal annotation* of the data. Annotating the data with a formal language serves, in this vision, two purposes: on the one hand, it results in "semantic" annotation: it records the *pure* meaning of the data, distilling it from the superstructures and the uncertainties of natural language; on the other hand, being formal, the language of the annotation allows one to make the same semantic assumptions that one does in standard data bases, namely that semantics can be formalized using methods similar to the formal semantics of programming languages. This semantic programme is based on the assumption that the semantic problems that one faces *vis à vis* data are not due to inherent characteristics of the data themselves, but to the defective way in which their meaning is carried by the language in which they are expressed. The foundational assumption is that all data have a *meaning*, which can be derived as a function of a certain representation of the data themselves. The problem is believed to be simply that these data are expressed in semiotic systems (from natural language to images and video) that problematizes the *extraction* of meaning. (The word "extraction" does a lot of work here: it provides the epistemological foundation of annotation, namely that meaning is an inherent quality of the data.) It is assumed, however, that meaning pre-exists the text (logically, if not chronologically), that can be expressed in a suitable formal system and associated with the data in a guise that can be understood by an algorithm.

The nodal points of information systems organized along these lines are the so-called *ontologies*, collections of axioms that intend to capture the semantics of the terms used in a certain domain of discourse and bring teh text that belong to the domain within the reach of standard, model-theoretic semantic approaches. Is ontology the correct way of representing meaning? By posing the problem in these terms one is begging the question of whether meaning can be represented at all, that is, whether it can be reified as a property of a document. Ontology says that it can, and that it can be represented as a collection of axioms on terms and relations. Since relations and their axioms are an important part of any ontology, an obvious way to start our analysis is to ask whether they are constitutive of meaning or not, that is, once we have represented a document by referring its elements to an ontology, whether the meaning resides in the terms themselves or in their relations.

Let us consider the first option first. This point of view is expressed quite well in Jerry Fodor's *informational semantics*:

> Informational semantics denies that "dog" means *dog* because of the way it is related to other linguistic expressions [...]. Correspondingly, informational semantics denies that the concept DOG has its content in virtue of its position in a network of conceptual relations[1].

---

[1] [3], p. 73.

The "correspondingly" here does a lot of work, and requires a fairly important metaphysical investment since it maps conceptual structures to linguistic ones. This, *passim,* is the same investment that ontology requires when it takes a linguistic structure (composed of words and relations) and calls it a conceptual model.

One of the problems of this point of view is that if one holds it as a theory of meaning, it is very hard to get out of radical nativism. That is, this model leads you almost automatically to admit that all concepts are innate, and almost none of them is acquired. This is quite absurd, of course: as Fodor says

[...] how could DOORKNOB be innate? *DOORKNOB*, of all things![2]

Fodor escapes this trap somehow, through the narrow door of assuming that a concept resides in how something *strikes us.* His way out relies heavily on a cognizant interpreter (somebody who can, at least, be *struck* by things), but this way Fodor has a hard time explaining the deep differences in the creation of concepts between different languages because it is not clear from it why should it be that the same thing strikes, say, Italian speakers differently that Japanese ones.

Let us get rid immediately of the idea that "dog" means DOG because of the three letters of which it is composed. There is absolutely nothing in the sequence /d/, /o/, and /g/ that is in any way connected to dogness. If you don't speak Italian, the sequence /c/, /a/, /n/, and /e/ doesn't mean anything to you, but to an italian it means more or less the same thing.

But if the letters themselves do not create any connection between the symbol "dog" and the meaning of the word, where does this connection come from? What is left of the symbol once you take away the elements that compose it? Where does its identity lie? The only way one can save the symbol is to say that its identity derives from its relations of opposition with the other symbols of the system. Dog is dog not because of the letters that make it up, but because they allow us to distinguish it from *dot*, from *hog*, from *god*. We are led, in other words, to a position that might oscillate between some form of cognitive functionalism [17] and structural semantics [5], depending on the degree to which we want to rely on logic formulas in order to define meaning. Both these positions, in spite of their fundamental differences, will agree that the meaning of a symbol is not in the symbol itself, but in the whole system, and in the relation of the symbols with the other symbols.

In mathematical terms, one can say that a system of signification must be invariant to any isomorphic transformation of its terms: if we change dog in hog, hog in bog, and so on, in such a way that the differences between symbols are maintained, the ontology that we get must be exactly equivalent to the original one. An isomorphism of this type will leave the relations between symbols unchanged so, if we take the second position outlined above—namely that the relations are constitutive of meaning—we obtain the necessary invariance. This position also entails that, whenever this relational invariance is not in force, meaning is not preserved. In other words: any transformations that is not an isomorphism of the terms of an ontology will not preserve meaning. A good way to test the plausibility of this assumption is to look at the relations between

---

[2] *ibid.* p. 123, emphasis in the original.

different languages. One can build many examples that show that languages are, indeed, far from isomorphic (a few ones can be found in [14]). Different languages can characterize the same semantic axis using different oppositions [5] or divide the semantic field using different semantic axes. To the extent that a functional translation from Chinese to English, or from Hungarian to Quechua is possible, then, we must admit that a meaning-preserving morphism doesn't have to be an isomorphism of terms that preserves relations[3]. Meaning, in other words, is a more abstract entity than a mere structural correspondence: depending on the global organization of the semantic field operated by a language, one can introduce considerable structural distortion and still end up with documents that can convey the same sense.

There is nothing structurally *in* the text that can be construed as meaning: meaning is not a property of the text, but a special kind of relation between the document and its interpreter. There is no meaning without interpretation, and interpretation is always contextually and historically situated.

2.1 Ontology as Non-contextual Meaning

The perspective on meaning given by ontology is very different from the contextual that is necessary in order to create meaning, and herein lies its main limitation. This limitation goes beyond the use of a specific logic system, and even beyond the limitations of any conceivable logic system: it derives from the disregard of interpretation as a creator of meaning and, consequently, from the idea that meaning is a *thing* rather than a process. Not only is the idea of formalizing meaning in a set of symbols and relations between them highly problematic, the very idea that the meaning of a document is *in* the document, that it can somehow be attached to the document in such a way that it can be revealed to a un-contextualized reading, is quite wrong.

An ontology encodes an absolute and immutable meaning of a text[4]. Where does it come from? For ontology to work, meaning must exist prior to text and independently of the language in which it is expressed. The scheme is pretty much that of a communication channel.

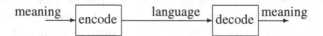

The origin of the communicative act is a *meaning* that resides with the author, and that the author wishes to express in a permanent text [6]. This meaning is a-historical, immutable, and pre-linguistic. In order to communicate meaning, the author translates it into the shared code of language, and sends it to the receiver. This translation may be imperfect; a contingency due to the accidental imperfections of human languages.

---

[3] As a matter of fact, it is not required to be a function at all: the idea of *one* correct translation has long disappeared from translation theory [12]. Rather, different translations are possible depending on the rôle that the translation will play in the receiving culture.

[4] This doesn't exclude the possibility that different encodings may give different, possibly conflicting, accounts of the meaning of a document, among which it may be necessary to negotiate. But every encoding will give one account of meaning, in absolute terms, that is, independently of the circumstances of interpretation.

Once meaning is translated into language, it can be delivered to the reader, who can then proceed to decode it obtaining a reasonable approximation of the original meaning as *intended* by the author.

This model of signification is necessary for the ontological enterprise because it is the only one that allows meaning to be *assigned* to a text, and recorded in a formal language other than the natural language, from which it can be extracted through automatic means following a schema like this:

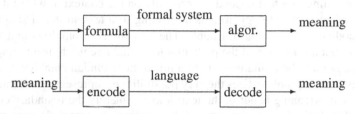

The conclusions of much of the linguistics and philosophy of language of the XX century, however, point in a different direction. There can be no meaning before language and independent of it: meaning can only exist within the categories and the strictures of language [7]. Not only meaning, but the signifying subject as well are a product of language [10]; there can be no pre-linguistic signification experience that belongs only to the author, because meaning can only be expressed in language, and language is a social construction.

It is the act of reading, contextual and situated, that gives a text its meaning. Reading is not a one-directional activity in which a reader is imbued with meaning; it is a dialectic process. It is an infinite process through which a frame of reference is created in which part of the text is interpreted, a text that changes the frame of reference and leads to a different interpretation of the text, which changes the frame of reference and so on... This process of framing and interpretation is what reception theorists call the *hermeneutic circle* [4,2].

Lest should you think that all this applies only to literature and not to the prosaic world in which ontology operates, let us consider an example that is often used in annotation: *a vegetarian pizza is a pizza with no meat*. This definition is quite clear as long as one can give a definition of "meat" (there is also some uncertainty on what one might consider a pizza, but let us ignore it here). This, in turn, depends in part on what kind of vegetarian one is: we go from people who eat fish to people who will not eat any animal product (vegans). If we go by the zoölogical definition of meat we are already in trouble because of the fish.

Then there is the dough. If the pizza was made from frozen dough it will probably contain saturated animal fats, which are added to make it crisp. Depending on the vegetarian, this will also count as meat, so most frozen foods (and therefore pretty much all restaurant food) will be out of the question. There may actually be in the pizza substances whose composition is the same as that of some molecule found in vegetables but that, for reasons of economy, were derived from an animal. Whether that counts as meat depends a lot on the reason why one is a vegetarian. If one is a vegetarian out of health reasons, these substances will probably be accepted; if one is as a protest against cruelty on animals, then the substance will be rejected. We could go on like this pretty much forever. The point is that in this analysis we have made reference, more than to the pizza itself, to the vegetarian who will (or won't) eat it, and to his relations to the pizza and to the practices of its production. You can't decide whether a pizza is vegetarian unless you do it from the point of view—cultural, social, and ideological—of the vegetarian: the receiver of the message. Any a priori classification would be a normative imposition of an ideology; in this case, it would be the seller of the pizza who decides what counts as vegetarian.

But if the meaning of a text depends so crucially on the context in which it is read, then the general plan of ontology, to attach meaning to a text so that a simple algorithm can decode is in quite a bit of trouble. The limitations of ontology that we have highlighted are not a limitation of the particular logic that one might use to implement it, nor of logic *per se*: the limitations are at a much more fundamental level. The discussion in this section problematizes the very possibility of representing meaning as an attribute of a text. Meaning is not *in* the text: a text is merely the boundary condition of a process that depends on the interpreter, his context, the linguistic community of which the interpreter is part, its discursive practices, etc. This doesn't necessarily imply that, for the purpose of meaning formation, the text can't be usefully represented using alternative means, including formal ones. As computing scientists, we are interested, pragmatically, in situations in which reading and interpretation are somehow mediated by a computer, and alternative representations of the text may favor this mediation. But in any case, whatever partial representation we have, we can never assume that we possess a representation of the meaning of the text, one from which meaning can be extracted in an a-contextual way by an algorithm.

3   Context-Based Retrieval

In the light of the previous observations, it seems clear that one can't hope to simply encode the semantics of a document in manner independent of the hermeneutic act of reading: meaning is created anew with each interpretation, and is a result of that operation. Our problems, then, are basically three: given a data access situation, we must (i) find a suitable context in which the data access is situated, (ii) find ways to formalize this context, at least to a certain degree (we are, after all, computing scientist, and we can only work with what we can formalize), and (iii) find ways in which the context can interact with the data to generate meaning.

Let us start with a fairly general theoretical model. We have said that the context in which a document is interpreted is essential to determine its meaning, that is, that the context *changes the meaning* of a text. We can also see things going in the opposite direction: the function of the semantics of a text is to *change the context* of the reader. If you are interested in literature, the context in which you look at American literature will not be the same after reading *Moby Dick*; if you travel on a motorway, your context will no longer be the same after seeing a speed limit sign. A document that doesn't change the context in which you act is, by definition, meaningless. We can express this situation with the following expression:

$$C_1 \xrightarrow{\mu(t)} C_2$$

where C_1 and C_2 are the contexts of the reader before and after interpreting the text, t is the text, and $\mu(t)$ is its meaning.

This is, as we have said, a very generic model, but we can use it to start answering some questions. For one thing, *is it possible to formalize meaning?* The answer of our model is that it is possible only to the extent that it is possible to formalize context. If C_1 and C_2 are formally defined in mathematical terms, then, and only then, it will be possible to give a formal definition of the function $\mu(t)$.

At one extremum, we have the situation in which the context can be completely formalized. This is the case, for instance, of programming languages: here the context can be reduced to the *state* of a computer on which the program is run, and the meaning of a program to a function that transforms an initial state of the computer to a final one. In other words, if the text is a program and the context of its interpretation is a computer system, meaning reduces to the usual denotational semantics of a program.

At the other extremum we have the general semiotic context, which we know can't be formalized in symbols, that is, given that a computer is a symbol manipulation machine, it can't be formalized in a computer.

The properties of the "space of contexts" depend crucially on the properties of the representation of the context that we have chosen, and it is therefore difficult to say something more about meaning is we don't impose some additional restriction. A reasonable one seems to be that we be capable of measuring the degree by which two contexts differ by means of an operation $\Delta(C_1, C_2) \geq 0$ such that, for each context C, it is $\Delta(C, C) = 0$. We don't require, for the time being, that Δ be a distance. Now the meaning of a document d in a context C can be defined as the difference that d causes to C:

$$\mu_C(d) = \Delta(\mu(d)(C), C) \tag{1}$$

Within this theoretical framework we can analyze, at least in the first approximation, various existing approaches, and devise ways to extend them. In this general scheme, the ontological approach to meaning can be synthesized as a constant function:

$$\bot \xrightarrow{\mu(d)} C \tag{2}$$

that is, ontology assigns a meaning to a document independently of the context in which the document is interpreted. This fact results, in our model, in the creation of a constant context, which depends only on the document and not on what was there before.

A very different point of view is that of *emergent semantics* [16,15]: in this approach, a highly interactive system allows the user and the system to organize the data in a way that highlights their contextual relations. The meaning of the data emerges as an epiphenomenon of this interaction. Emergent semantics does not work with one document at the time, but always with set of documents, since meaning always emerges from relations. Therefore, the meaning function μ will take as argument a suitable configuration D of documents. The user action is represented as an operator u, and the schema is the following:

$$C \underset{u}{\overset{\mu(D)}{\rightleftarrows}} C' \tag{3}$$

The context oscillates between C, which is the new contextual situation in which the user wants to end, and C', which is the context proposed by the computer with the access to the new documents. The semantic function is, in this case, the equilibrium of the cycle or, in other terms, the least fix-point of the function $\mu(D) \circ u$.

The model that we have outlined in the previous section entails the demise of search and querying as identifiable and independent activities. The "death of the query" is the price that we have to pay for semantics, for if semantics can only be present in the context of a certain activity, then search can only be conceived as part of that activity, possibly as something of a very different nature for each different activity. In this analysis of the transformation of querying we receive some help from the Wittgensteinian notion of *Sprachspiel* (language game). Wittgenstein purposely didn't define exactly what a Sprachspiel was, on the ground that the different games are not related by a fixed set of criteria but by a "family resemblance" [18]. We can say, with a certain degree of approximation typical of a formalizing discipline like computing science, that a Sprachspiel is a linguistic activity coördinated, at least partially, by a number of norms (some of which are implicit) that determine which language acts (the *moves* of the game) are permissible and which are their effects on the context of the game.

From this vantage point, what used to be called a query is not an activity but a type of move in a computing-linguistic game.

4   Implementing Context

The practical problems posed by the general orientation presented here include how to capture ongoing activities, how to represent them and, to the extent that it's possible, formalize them, in such a way that they can be used as a basis for data access. In general, of course, this is impossible. If a person is, say, shopping for detergent and wants to search the internet for brands with certain characteristics, there is very little hope that we can represent the activity "shopping for detergent" in a computer system: we are in this case in the presence of a physical activity that leaves no *digital trace*, so to speak.

On the other hand, a significant number of daily activities are, for many of us, executed on or with the aid of a computer, and they do have a digital trace, one that can be recorded and used as a context for a language game carried out as part of that activity. Suppose that we are preparing a presentation for a conference to which we had submitted a paper and that, during this process, we need to clarify a point or to look for an illustration for the presentation. In order to prepare the presentation, we have created a document in a directory (let us say the directory *presentation*) where we have possibly copied some documents that we thought might be useful. This directory is likely to be placed in a hierarchy as in figure 1. Its sibling directories will contain documents somehow related to the topic at hand although, probably, not so directly as those that can be found in the work directory. The siblings of the conference directory (and their descendants) will contain documents related to my general area of activity, although not necessarily directly related to the topic of the presentation. This information, suitably encoded, will constitute the context of the game. In order to create and play it, we have to specify two things: how to represent the context and how the various moves that the game allows will modify it; in particular, in this example, how the query moves of the game modify it.

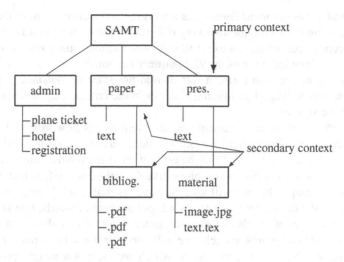

Fig. 1. The structure of directories and context for the preparation of a presentation

4.1 Context Representation

In order to build a representation, we consider two types of contexts: the *primary context* consists of the directory in which the current activity is taking place; the *accessory context* consists of any other directory that contains material in some capacity related to the current activity. The accessory context contains, in general, the descendants of the work directory and, possibly, its parent. This choice is somewhat *ad hoc*, and it is foreseeable that different systems will choose to use different *context policies* in order to determine the accessory context.

In each directory we create first a representation that takes into account only the files contained therein; we call such representation the *generator* of the directory. Then, for each directory, we create a further representation, called the *index*, built based on the generator of the directory itself (viz. of the primary context of the activities that take place there) and of the accessory contexts, as per the specific context policy adopted.

In the above example, in each of the six directories a generator will be created, with an appropriate representation of the context of that directory (that is to say, a representation of the documents that appear in the directory). The generators of the *pres* directory (the primary context) and of the directories *paper*, *bibliog.* and *material* (the accessory context), will join using appropriate operators, to form the index of the context of the search, which is stored in the directory *pres*. It must be noted that the construction of the index through generators supposes a hypothesis of compositionality of the context representation: the representation of the global context of two or more directories depends only on the representations of the local contexts and the relation between directories.

Let us begin by considering the construction of a generator, that is, of the context of a single directory that depends only on the documents found in the directory. In this example, we represent contexts using a technique similar to that of the semantic map WEBSOM [9]. This semantic map presents two features that are essential in our case: the representation of context by means of *self-organizing maps* in the Euclidean space of words, and the use of *word contexts* as a working and learning unit of the map. Note that we are using the technique in a very different capacity than that for which it was originally conceived: we do not use it to represent the data space but the context; that is, its function is not indexing as in [9], but query formation.

The *self-organizing* map forms a sort of non-linear *latent semantic* [1] space, and this non-linearity will is when making changes in the context (e.g. to express a query, as we shall see shortly).

Many representations of documents use the frequencies of words of the document; this representation is insufficient for our problem because if we use only a word by itself, the semantics that derives from the colocation of the words, namely the semantic component that is needed to solve problems like the polysemy, will be lost. On the other hand, in the technique that we will use, the fundamental unit of representation that is extracted from the document is not the word, but a group of words, that is called *word context*. The number of words of the *word context* may vary, in this work we consider the simplest case: two words, namely, we will consider pairs. Each pair of consecutive words in the text is seen as a symbol to which we assigns a weight proportional to the number of times the symbol (in other words, the pair of words) appears in the text (fig. 2 left).

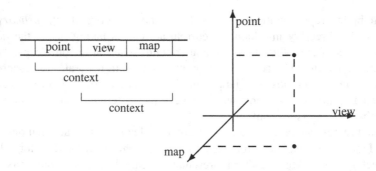

Fig. 2. The geometry of the words context

These pairs are represented in the typical geometric space of many information retrieval systems, a space in which each word is an axis. Since our basis are the contexts, the points in this space are not points in one of the axes (as in the case of simple words: each point is a word with its weight), but points in two-dimensional sub-spaces: each pair is a point in the plane represented by the two words that compose it. Using more complex contexts will result in points contained in spaces of higher dimension. As customary, before considering the words for the construction of indices, we will perform stop-word removal and stemming.

The *index* is a union of the generators of the primary and accessory contexts. In the case of our reference activity, the accessory context is composed of the descendants and the parent of the work directory. The weight of the pair constitute by the word number i and word number j (in other words, the word pair who has values in the e_i and e_j axes of the space of words), which may appear in several directories of the work context, is

ω^{ij}. Each generator that we use in order to compute the context has its own weight for the pair, assigned depending on the frequency of that pair in the local directory. Let ω_P^{ij} be the weight for the pair i, j in the primary context folder, S_k be the kth directory that composes the accessory context ($k = 1, \ldots, S$), and ω_k^{ij} the weight in that directory. Then the weight of the pair i, j in the context, ω^{ij} is given by the weighted linear combination:

$$\omega^{ij} = \gamma \omega_P^{ij} + \frac{1 - \gamma}{S} \sum_k \omega_k^{ij} \tag{4}$$

where γ is a constant, $0 \leq \gamma \leq 1$.

The map consists of a matrix of $N \times M$ neurons, each neuron being a vector in the word space; if the context is composed of T words, the neuron μ, ν ($1 \leq \mu \leq N$, $1 \leq \nu \leq M$) is a vector

$$[\mu\nu] = (u_{\mu\nu}^1, \ldots, u_{\mu\nu}^T) \tag{5}$$

The map learning is being developed under the stimulus of a set of points in input space, each point representing a pair of words *(word context)*. Given a total number of P pairs, and given that pair number k consists of the words number i and j, the corresponding point in the input space is given by

$$p_k = (\overbrace{0, \ldots, \omega^{ij}, 0, \ldots, \underbrace{\omega^{ij}}, 0, \ldots, 0}^{i}) \tag{6}$$

where ω^{ij} is the weight of the pair of words determined as in (4). During learning the p_k vectors are presented several times to the map. We call *event* the presentation of a vector p_k, and *iteration* the presentation of all vectors. Learning consists of several iterations. An event in which the vector p_k is presented entails the following operations:

i) Identify the "winning" neuron, in other words the neuron that is closer to the vector p_k:

$$[*] = \min_{[\mu\nu]} \sum_{j=1}^{T} (p_k^j u_{\mu\nu}^j)^2 \tag{7}$$

ii) The winning neuron, $[*]$, and a certain number of neurons in its "neighborhood" are moving toward the p_k point an amount that depends on the distance between the neuron and the winner one and the number of iterations that have been performed so far. For it, we define the *distance* between the neurons of the map as:

$$\|[\mu\nu] - [\mu'\nu']\| = |\mu - \mu'| + |\nu + \nu'|, \tag{8}$$

for $t = 0, 1, \ldots$ the counter of the iterations of the learning. We define a function of environment $h(t, n)$ such that

$$\forall t, n \geq 0 \; 0 \leq h(t, n) \leq 1, h(t, 0) = 1$$
$$h(t, n) \geq h(t, n + 1) \tag{9}$$
$$h(t, n) \geq h(t + 1, n)$$

and a coefficient of learning $\alpha(t)$ such that

$$\forall t \geq 0, 0 \leq \alpha(t) \leq 1, \alpha(t) \geq \alpha(t+1) \tag{10}$$

Then each neuron $[\mu\nu]$ of the map moves toward the point p_k according to the learning equation

$$[\mu\nu] \leftarrow [\mu\nu] + \alpha(t)h(t, \|[*] - [\mu\nu]\|)(p_k - [\mu\nu]) \tag{11}$$

The function h generically corresponds to an environment of the winning neuron that is done smaller as it increases the number of iterations. In this work the environment function is the Gaussian $h(t, n) = \exp(-n^2/\sigma(t)^2)$, con $\sigma(t) \geq \sigma(t+1) > 0$.

At the end of the learning process the map is laid out in the space of a word in a way that, in the extreme case of an infinite number of neurons that form a continuum, it optimally approximates the distribution of the points in the space [13]. This map represents the semantic space of the context and, as we mentioned in the previous section, can be assimilated to a nonlinear form of latent semantics.

4.2 The Query

In its most complete and general form, the procedure of a query is composed of four phases:

i) through an appropriate user interface or with a program that the user is using, an initial specification of the query is collected, we will name it the *proto-query*. The proto-query can be formed by a few words typed by the user, a paragraph that the user is editing, etc.. In a multimedia system the proto-query also contain an indication of the type of the document that's being searched (text, image, video, etc.)..

ii) The proto-query is used to change the current context, transforming it into a *objective context*. In practice, the configuration of the map (index) of the current directory is modified through a partial learning, which will give the context a *bias* towards the proto-query. The resulting configuration from this learning could be considered, in some way, as the interpretation of the proto-query in the actual context.

iii) The difference between the actual and objective context is the *differential context* and, in our model of semantics, corresponds to the semantic of the ideal document that is searched for: the document that is assimilated to the current context, will transform it into the objective context. An opportune codification of the ideal document is created and sent to the search server to retrieve the documents that more respond to that profile.

iv) The documents elected (e.g. read or downloaded) become part of the context: a new learning is run so that the current context reflects the new situation.

This general model of a query assumes the existence of a search service *(search engine)* capable of managing them. The construction of such a service is one of future goals of our work. For the moment, our objective is to demonstrate the role played by the context using it to focus searches on existing services. Therefore, it is necessary to transform

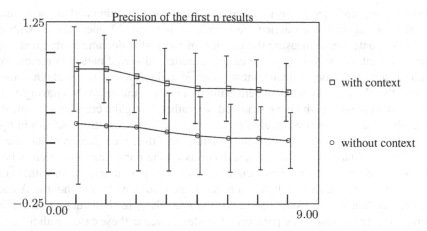

Fig. 3. Precision of the results, with and without context

the differential context into a list of words with weights, because the search services only accepts (if accepts) this type of queries. Obviously this type of query can not make an optimal use of the possibilities of context but, we repeat it, at this moment our goal is simply to evaluate the influence of the use of the context in the search. In our tests, the proto-query P is a set of keywords u_i. A keyword that correspond to the i word of the space is represented as the vector $e_i = (\overbrace{0, \ldots, 1}^{i}, 0, \ldots, 0)$. For simplicity we assume that every word in the query has the same weight w_i. Therefore, the query Q, formed by q words, will be represented as a point in the T-dimensional space: $Q = w \sum_{u_i \in P}^{q} e_i$

This vector is used for a partial learning process using the algorithm presented. During this process the neuron $[\mu\nu]$ is moved to the position $[\mu'\nu']$. The differential context is given by the differences of the neurons positions, $\delta_{\mu\nu} = [\mu'\nu'] - [\mu\nu]$ for each $[\mu\nu]$ in a neighborhood of the winning neuron (the closest neuron to the vector Q).

Projecting the vector $\delta_{\mu\nu}$ on the axes of the words, we get the weights of the words given by this neuron: $\delta_{\mu\nu} = (v_{\mu\nu}^1, \ldots, v_{\mu\nu}^T)$. The *non-normalized weight* of the word i is given by the sum of their weights relative to all the neurons in a neighborhood A of the winning neuron

$$V^i = \sum_{[\mu\nu] \in A} v_{\mu\nu}^i \tag{12}$$

Considering only the K words with greater weights, and normalizing the vector of weights for these words we obtain the query that will be send to the search engine, composed of a set of words each one associated with a weight.

Testing fully the context approach is quite problematic at this time for lack of a proper contextual server and its data base infrastructure. In order to obtain some preliminary indications, we used the limited weighting capabilities offered by the *google* commercial search engine (www.google.com). The contextual query was translated in a collection of weighted terms, and weighting was roughly approximated through positioning and repetition in the search engine query. As context, we considered, for the example reported here, the directory structure in the computer of one of us (Santini), and as working directory a directory with several columns by that author for the magazine *IEEE Computer*. We queried the search engine with 32 query terms, with and without the context, and measure the fraction of the first n documents that were considered relevant, for $1 \leq n \leq 8$. Given the generic and varied nature of the columns contained in the directory, a document was considered relevant if it was about computing. Note that the measure adopted here is the precision of the result. Not having a fixed corpus of documents in which we searched (or, rather, being the corpus the whole data base of the search engine) we couldn't measure recall. The results are shown in fig. 3 It is evident even without a detailed analysis that the difference is large and statistically significant. Qualitatively, the difference depends on the particular query that is being made. Very technical words, whose semantic span is very limited to begin with, benefit little from the context, and fetch basically the same results with or without it. A query word such as "algorithm", for instance, is quite unlikely to fetch documents not related to computing, regardless of the presence of context. Even in these cases, with the use of context, there seemed to be a better "focus" of the results fetched around topics related to the columns, but we have devised to way to quantify this effect. On the opposite side, queries with ambiguous terms, such as "sort" (data sort in computing, an approximation of qualities in the common language) gave the most dramatic improvements when context was used.

5   Words of Parting

We have argued that formal annotation, and the general ontological programme that comes with it, might not be the proper way to consider the problem of the meaning of the document and, in general, to frame the issues related to semantics. This is not a majority opinion, not by a long shot, and there are a few reasons that contribute to its unpopularity and to the exclusivity of the attention given to annotation and ontology.

First, there is the pull of certain common sense philosophy. We can look at texts, read them, and make sense of them, and it seems natural to interpret this act as unlocking the meaning that is *in* the text. After all, if we don't know which gate does flight 354 to New York leave from, and we read the announcement board of the airport, we end up knowing it. It is easy to model this situation as a transfer of a specific information (viz. that the flight leaves from gate C34) from the announcement board to the reader. The error is the failure to recognize that the model can be construed to approximately work only in a limit case like this one, namely a case in which the external context is so constraining that the appearance of the symbol "C34" can basically have only one interpretation, and to extend the same model to the common situation, the one in which the interpretation context plays a much more important rôle. We have given arguments why this position represents a gross philosophical simplification, and we believe that it will ultimately result in the sterility of semantic computing.

Second, there is the understandable inertia of an established position on which a considerable intellectual and financial investment has been made. The agencies and companies that have invested in annotation and ontology are obviously more eager to see their approach produce results than in exploring alternatives. This phenomenon is quite well understood in the modern epistemological literature [8].

Finally, there is a point related to the economy of the commercial web (which, unlike ten years ago, today represents the vast majority of the web today). The model of meaning assumed by the semantic web is very appealing to web companies because, if meaning is inherent in a text, it can be owned, bought, and sold like other goods. Lyotard, in 1979, observed a similar phenomenon regarding knowledge: "knowledge is and will be produced in order to be sold, is and will be consumed in order to be valued in production: in both cases, in order to be exchanged"[5]. Lyotard considers this phenomenon as a natural consequence of the computerization of knowledge: "[knowledge] can go through the new channels [(those of informatics)] and become operational only if it can be translated in amount of information"[6]. It is not too daring, then, to expect that a similar change will occur with respect to meaning once this has been codified in formal annotations: only meaning that *can* be codified will survive, and this will do so only in order to be exchanged as merchandise.

In the ontology view, meaning is a property of the author (or of the organizations that bought it from the author), a property that can be exchanged with the reader using the currency of language. Among other things, this "market" view of meaning opens the logical possibility of copyrighting meaning, patenting meaning, and in general posing commercial restrictions to the free exchange of meaning. For those of us who believe that the web should be a common good, in which commercial interests should never replace the free exchange of ideas, this is not an appealing perspective.

Technically, this paper has presented the outline of a different model of meaning, one in which the reader's context plays a preponderant rôle. We have presented a simple framework in which we are currently experimenting with this model, a framework that in the future will be extended in different directions: on the one hand, the integration in this framework of more formal representations, at least for those parts of the context that can be formalized; on the other hand, the development of suitable data base techniques to make this kind of query efficient.

Our purpose will be, on the one hand, to build a context-based data access client (configured as a plug-in to some word processing or presentation program, if possible) to make context based retrieval on general web sites and repositories and, on the other hand, to build a context-based access server. The latter will be akin to the servers built for search engines such as yahoo or google but, while these servers do not coöperate with the user's computer (apart from the elementary communication necessary to retrieve the query and return the results), the server that we consider here will be integrated with the user's computer from which it will derive the current context, and with which it will coöperate to support interaction.

References

1. Deerwester, S., Dumais, S.T., Furnas, G.W., Landauer, T.K., Harshman, R.: Indexing by latent semantic analysis. Journal of the American Society for Information Science 41(6), 391–407 (2000)
2. Eco, U.: The role of the reader. University of Indiana Press, Bloomington (1979)

---

[5] [11], p. 14, our translation.
[6] *ibid.* p. 13

3. Fodor, J.: Concepts. Oxford University Press, Oxford (1997)
4. Gadamer, H.-G.: Truth and method. London (1975)
5. Greimas, A.J.: Sémantique structurale. larousse, Paris (1966)
6. Hirsch, E.D.: Validity in interpretation. New Haven, Conn. (1967)
7. Jameson, F.: The prison-house of language. Princeton, NJ (1972)
8. Juhn, T.: The structure of scientific revolutions. Chicago University Press, Chigago (1996)
9. Kaski, S.: Computationally efficient approximation of a probabilistic model for document representation in the WEBSOM full-text analysis method. Neural Processing letters 5(2) (1997)
10. Lacan, J.: Escrits: a selection. Norton, New York (1982)
11. Lyotard, J.-F.: La condition postmodérne. Editions de minuit, Paris (2001)
12. Moya, V.: La selva de la traducción. Cátedra, Madrid (2004)
13. Santini, S.: The self-organizing field. IEEE Transactions on Neural Networks 7(6), 1415–1423 (1996)
14. Santini, S.: Ontology: use and abuse. In: Boujemaa, N., Detyniecki, M., Nürnberger, A. (eds.) AMR 2007. LNCS, vol. 4918. Springer, Heidelberg (2008)
15. Santini, S., Gupta, A., Jain, R.: Emergent semantics through interaction in image databases. IEEE Transaction on Knowledge and Data Engineering (2001)
16. Santini, S., Jain, R.: The "el niño" image database system. In: Proceedings of the IEEE International Conference on Multimedia Computing and Systems, pp. 524–529 (1999)
17. Stich, S.: From folk psychology to cognitive science. MIT Press, Cambridge (1983)
18. Wittgenstein, L.: Philosophical Investigations. Prentice Hall, Englewood Cliffs (1973)

Exploring Music Artists via Descriptive Terms and Multimedia Content

Markus Schedl and Tim Pohle

Department of Computational Perception
Johannes Kepler University
Linz, Austria
music@jku.at
http://www.cp.jku.at

Abstract. This paper presents an approach to browse collections of web pages about music artists by means of descriptive terms and multimedia content. To this end, a user interface called *Three-Dimensional Co-Occurrence Browser (3D-COB)* is introduced. 3D-COB automatically extracts and weights terms from artist-related web pages. This textual information is complemented with information on the multimedia content found on the web pages. For the user interface of 3D-COB, we elaborated a three-dimensional extension of the Sunburst visualization technique. The hierarchical data to be visualized is obtained by analyzing the web pages for combinations of co-occurring terms that are highly ranked by a term weighting function.

As for evaluation, we investigated different term weighting strategies in a first user study. A second user study was carried out to assess ergonomic aspects of 3D-COB, especially its usefulness for gaining a quick overview of a set of web pages and for efficiently browsing within this set.

1 Introduction

Automatically finding descriptive terms for a given music artist is an important task in music information retrieval (MIR). Such terms may describe, for example, the genre or style of the music performed by the artist under consideration and enable a wide variety of applications, e.g., enriching music players [19], recommending unknown artists based on the user's favorite artists [24], or enhancing user interfaces for browsing music collections [13,17,10,15,22].

One possibility for assigning musically relevant terms to a given artist is manual annotation by music experts or communities, as it is usually employed by music information systems like *allmusic* [5] and *last.fm* [4] or interfaces for music search like *musiclens* [3]. However, this is a very labor-intensive task and barely feasible for huge music collections. An alternative way, which we follow here, is to exploit today's largest information source, the World Wide Web. Automatically deriving information about music artists from the web is advantageous since it incorporates the opinions of a large number of different people, and thus embodies a kind of cultural knowledge.

D. Duke et al. (Eds.): SAMT 2008, LNCS 5392, pp. 137–148, 2008.

The *Three-Dimensional Co-Occurrence Browser (3D-COB)* presented here automatically indexes a set of web pages about music artists according to a dictionary of musically relevant terms and organizes these web pages by creating a number of subsets, each of which is described by a set of terms. The terms that describe a particular subset are determined by a term weighting function. The subsets are then visualized using a variant of the Sunburst technique [8,21].

The purpose of 3D-COB is threefold. First, it facilitates getting an *overview* of the web pages related to a music artist by structuring them according to co-occurring terms. Second, since the descriptive terms that most often occur on web pages related to a music artist X constitute an individual profile of X, 3D-COB is also suited to reveal various *meta-information* about the artist, e.g., musical style, related artists, or instrumentation. Third, by visualizing the amount of *multimedia content* provided at the indexed web pages, the user is offered a means of exploring the audio, image, and video content of the respective set of web pages.

2   Related Work

This paper is mainly related to the two research fields of *web-based music information retrieval* and *information visualization of hierarchical data*, which will be covered in the following.

Determining terms related to a music artist via web-based MIR has first been addressed in [23], where Whitman and Lawrence extract different term sets (e.g., noun phrases and adjectives) from artist-related web pages. Based on term occurrences, individual term profiles are created for each artist. The authors then use the overlap between the term profiles of two artists as an estimate for their similarity. A quite similar approach is presented in [12]. Knees et al. however do not use specific term sets, but create a term list directly from the retrieved web pages. To this list, a term selection technique is applied to filter out less important terms. Hereafter, the TF·IDF measure is used to weight the remaining words and subsequently create a weighted term profile for each artist. Knees et al. propose their approach for artist-to-genre classification and similarity measurement. Both approaches presented so far barely address the potential of utilizing the large amount of terms which, though not directly related to the artist, occur on many artist-related web pages. In contrast, Pampalk et al. in [16] use a dictionary of about 1,400 musically relevant terms to index artist-related web pages. Different term weighting techniques are applied to describe each artist with some terms. Furthermore, the artists are hierarchically structured using a special version of the Self-Organizing Map. The authors show that considering only the terms in the dictionary for term weighting and clustering outperforms using all terms found on the extracted web pages. An approach to assign descriptive terms to a given artist is presented in [19]. Schedl et al. use co-occurrences derived from artist-related web pages to estimate the conditional probability for the artist name under consideration to be found on a web page containing a specific descriptive term and vice versa. To this end, a set of predefined genres and other attributes, like preferred tempo or mood of the artist's performance, is used. The aforementioned probabilities are then calculated, and the most probable value of the attribute under consideration is assigned to the artist. Independent of Schedl et al., Geleijnse and Korst present in [9] an approach that differs from [19] only in regard to the normalization used.

The 3D-COB proposed here uses a dictionary like [16] to extract artist-related information from web pages. However, the clustering is performed in a very different way and on a different level (for individual web pages instead of artists).

Related work on visualizing hierarchical data primarily focuses on the *Sunburst* approach, as the 3D-COB extends the Sunburst in various aspects. The Sunburst as proposed in [8,21] is a circular, space-filling visualization technique. The center of the visualization represents the highest element in the hierarchy, whereas elements on deeper levels are illustrated by arcs further away from the center. Child elements are drawn within the angular borders of their parent, but at a more distant position from the center. In almost all publications related to the Sunburst, its usual application scenario is browsing the hierarchical tree structure of a file system. In this scenario, directories and files are represented by arcs whose sizes are proportional to the sizes of the respective directories/files. In the case of the 3D-COB, however, some *constraints for the size of the visualization* are necessary. Furthermore, the arc sizes are determined by the *term weighting function*, which is applied to select the most important terms (for the clustering). Moreover, the 3D-COB allows for *encoding an additional data dimension* in the height of each arc. This dimension is used to visualize the amount of multimedia content provided by the analyzed web pages. As three different types of multimedia content are taken into account (audio, image, and video), the Sunburst stack of the 3D-COB consists of three individual Sunburst visualizations.

Other space-filling visualization techniques for hierarchically structured data include the *Treemap* [11]. In contrast to the Sunburst, the Treemap uses a rectangular layout and displays elements further down in the hierarchy embedded in the rectangle of their parent element. The Sunburst, however, displays all elements that reside on the same hierarchy level on the same torus, which facilitates getting a quick overview of the hierarchy.

3   The Three-Dimensional Co-occurrence Browser (3D-COB)

To get a first impression of the appearance of the 3D-COB user interface, the reader is invited to take a look at Figure 1. This figure shows a stack of three three-dimensional Sunburst visualizations created from 161 web pages of the band "Iron Maiden". Details on the information gathering process, the creation of the visualization, and the user interaction possibilities are provided in the following subsections.

3D-COB has been implemented using the *processing* environment [6] and the *CoMIRVA* framework [18]. CoMIRVA already contained an implementation of the two-dimensional version of the Sunburst. We heavily extended this version by heaving it to the third dimension and incorporating various multimedia content in the visualization. To this end, we particularly had to elaborate new data representation and user interaction models.

3.1   Retrieval and Indexing

Given the name of an artist, we first query *Google* with the scheme *"artist name" +music +review* to obtain the URLs of up to 1,000 web pages related to the artist, whose content we then retrieve. Subsequently, a term analysis step is performed. To this end, we use a dictionary of musically relevant terms, which are searched in all web pages of every artist, yielding an inverted file index. For the conducted experiments, a manually compiled dictionary that resembles the one used in [16] was utilized. It was assembled using various sources such as [7,5,1] and contains music genres and styles, instruments, moods, and other terms which are somehow related to music. Altogether, the dictionary contains 1,506 terms.

As for indexing the multimedia content of the web pages, we first extract a list of common file extensions for audio, image, and video files from [2]. We then search the HTML code of each web page for links to files whose file extension occur in one of the extracted lists. Finally, we store the URLs of the found multimedia files and the inverted file index gained by the term analysis in an XML data structure.

3.2   Creation of the Visualization

Using the inverted index of the web pages of an artist X, we can easily extract subsets $S_{X,\{t_1,\ldots,t_r\}}$ of the web page collection of X which have in common the occurrence of all terms $t_1, \ldots, t_r$.

Starting with the entire set of web pages $S_{X,\{\}}$ of X, we use a term weighting function (e.g., document frequency, term frequency, TF·IDF) to select a maximum number N of terms with highest weight, which are used to create N subsets $S_{X,\{t_1\}}, \ldots, S_{X,\{t_N\}}$ of the collection. These subsets are visualized as arcs $A_{X,\{t_1\}}, \ldots, A_{X,\{t_N\}}$ around a centered cylinder which represents the root arc $A_{X,\{\}}$, and thus the entire set of web pages. The angular extent of each arc is proportional to the weight of the associated term t_i, e.g., to the number of web pages containing t_i when using document frequencies for term weighting. To avoid very small, thus hardly perceivable, arcs, we omit arcs whose angular extent is smaller than a fixed threshold E, measured in degrees. Furthermore, each arc is colored with respect to the relative weight of its corresponding term t_i (relative to the maximum weight among all terms). The term selection and the corresponding visualization step are recursively performed for all arcs, with a maximum R for the recursion depth. This eventually yields a complete Sunburst visualization, where each arc at a specific recursion depth r represents a set of web pages $S_{X,\{t_1,\ldots,t_r\}}$ in which all terms $t_1, \ldots, t_r$ co-occur.

As for representing the multimedia content found on the web pages, in each layer of the Sunburst stack, the amount of a specific category of multimedia files is depicted. To this end, we encode the relative number of audio, image, and video files in the height of the arcs (relative to the total number represented by the root node of the respective layer). For example, denoting the audio layer as L_A, the image layer as L_I, and the video layer as L_V and focusing on a fixed arc A, the height of A in L_I shows the relative number of image files contained in the web pages that are represented by arc A, the height of A in L_V illustrates the relative number of video files, and the height of A in L_A the relative number of audio files. The corresponding multimedia files can easily be accessed via the user interface of 3D-COB.

3.3   User Interface and User Interaction

Figure 1 depicts a screenshot of 3D-COB's user interface for 161 web pages retrieved for "Iron Maiden". The constraints were set to the following values: $N = 6$, $R = 8$, $E = 5.0$ (cf. Subsection 3.2). Document frequencies were used for term weighting. Each arc $A_{X,\{t_1, ..., t_r\}}$ is labeled with the term t_r that subdivides the web pages represented by the arc's parent node $A_{X,\{t_1, ..., t_{r-1}\}}$ into those containing t_r and those not containing t_r. Additionally, the weight of the term t_r is added in parentheses to the label of each arc $A_{X,\{t_1, ..., t_r\}}$. The topmost layer illustrates the amount of video files found on the web pages, the middle one the amount of image files, and the lower one the amount of audio files. In the screenshot shown in Figure 1, the arc representing the web pages on which all of the terms "Iron Maiden", "guitar", and "metal" co-occur is selected. Since document frequencies were used for this screenshot, determining the exact number of web pages represented by a particular arc is easy: 74 out of the complete set of 161 web pages contain the mentioned terms.

User interaction is provided in several ways. First, the mouse can be used to *rotate* the Sunburst stack around the Y-axis, i.e., the vertical axis going through the root nodes of all Sunbursts in the stack. *Zooming* in/out (within predefined boundaries) is provided as well as *changing the inclination of the stack*, which is limited to angles between a front view and a bird's eye view. To select a particular arc, e.g., to access the multimedia content of the corresponding web pages, the arrow keys can be used to navigate in the hierarchy. The currently selected arc is highlighted by means of drawing a white border around it and coloring its label in white. So are all previously selected arcs at higher hierarchy levels. This facilitates tracing the selection back to the root arc and quickly recognizing all co-occurring terms on the web pages represented by the selected arc.

In addition to the basic interaction capabilities described so far, the following functionalities are provided.

- Creating a new visualization based on the subset of web pages given by the selected arc.
- Restoring the original visualization that incorporates all web pages in its root node.
- Showing a list of web page URLs which are represented by the selected arc.

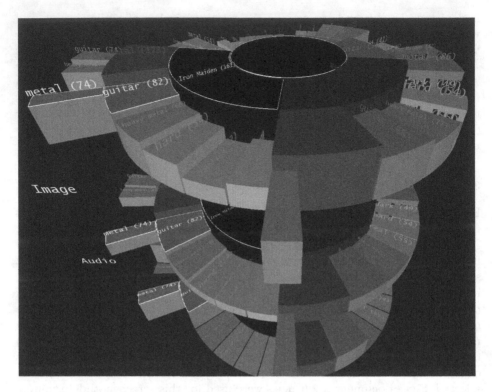

Fig. 1. The user interface of 3D-COB for a collection of web pages about the band "Iron Maiden"

– Displaying and browsing a list of audio, image, or video files, which are found on the web pages of the currently selected arc.
– Opening the web pages or the available multimedia files represented by the selected arc.

4 Evaluation of the Term Weighting Functions

We experimented with three different term weighting functions (document frequency, term frequency, TF·IDF) for term selection in the Sunburst creation step, cf. Subsection 3.2. Given a set of web pages S of an artist, the *document frequency* DF_t of a term t is defined as the absolute number of pages in S on which t appears at least once. The *term frequency* TF_t of a term t is defined as the sum of all occurrences of t in S. The *term frequency inverse document frequency* measure $TF \cdot IDF_t$ of t is calculated as $TF_t \cdot \ln \frac{|S|}{DF_t}$.

To assess the influence of the term weighting function on the quality of the hierarchical clustering, the hierarchical layout, and thus on the visualization of 3D-COB, we conducted a user study as detailed in the following.

4.1 Setup

For the user study, we chose a collection of 112 well-known artists (14 genres, 8 artists each). Indexing was performed as described in Subsection 3.1. To create the evaluation data, for each artist, we calculated on the complete set of his/her retrieved and indexed web pages, the 10 most important terms using each of the three term weighting functions. To avoid biassing of the results, we combined, for each artist, the 10 terms obtained by applying every weighting function. Hence, every participant was presented a list of 112 artist names and, for each of the name, a set of associated terms (as a mixture of the terms obtained by the three weighting functions). Since the authors had no a priori knowledge of which artists were known by which participant, the participants were told to evaluate only those artists they were familiar with. Their task was then to rate the associated terms with respect to their appropriateness for describing the artist or his/her music. To this end, they had to associate every term to one of the three classes + (good description), – (bad description), and ~ (indifferent or not wrong, but not a description specific for the artist).

Due to time constraints, we had to limit the number of participants in the user study to five. Three of them are computer science students, the other two researchers in computer science. All of them are male and all stated to listen to music often.

4.2 Results and Discussion

We received a total of 172 assessments for sets of terms assigned to a specific artist. 92 out of the 112 artists were covered. To analyze the results, we calculated, for each artist and weighting function, the sum of all points obtained by the assessments. As for the mapping of classes to points, each term in class + contributes 1 point, each term in class – gives -1 point, and each term in class ~ yields 0 points.

The complete results for each artist assessed cannot be depicted here due to space limitations.[1] However, Table 1 shows the sum of the points (over all artists) for each term weighting function as well as the average score obtained for an arbitrarily chosen artist. Investigating the averaged points for different artists (which are not shown in the table) reveals that the quality of the terms vary strongly between the artists. Nevertheless, we can state that, for most artists, the number of descriptive terms by far exceeds that of the non-descriptive ones. Due to the performed mapping from classes to points, the averaged score values can be regarded as the average excess of the number of good terms over the number of bad terms. Hence, overall, we assume that the document frequency measure performed best, the term frequency second best, and the TF·IDF worst.

To test for the significance of the results, we performed Friedman's non-parametric two-way analysis of variance [20]. This test is similar to the two-way ANOVA, but does not assume a normal distribution of the data. The test yielded

---

[1] Detailed results per artist can be found at [omitted due to blind review].

Table 1. For each term weighting function, the summed up user ratings and the average ratings per artist

	TF	DF	TF·IDF
sum	386.00	413.00	271.00
avg	2.22	2.44	1.53

a p of 0.000024. Therefore, it is highly probable that the variance differences in the results are significant. Moreover, pairwise comparisons between the results given by the three term weighting functions showed that TF·IDF performed significantly worse than both TF and DF, whereas no significant difference could be made out between the results obtained using DF and those obtained using TF.

The laborious task of combining and analyzing the different assessments of the participants in the user study allowed us to take a qualitative look at the terms. Overall, the majority of the terms was judged descriptive. However, we discovered an interesting source of error: the erroneous assignment of a term to an artist if the term is part of artist, album, or song name. Examples for this problem are "infinite" for the artist "Smashing Pumpkins" as well as "human" and "punk" for the artist "Daft Punk".

5   Evaluating the User Interface

To investigate the usefulness of 3D-COB for gaining a quick overview of a set of artist-related web pages and efficiently browsing within this set, we conducted a second user study that primarily focuses on ergonomic aspects of 3D-COB.

5.1   Setup

We formulated the following tasks, which we believe are important for the mentioned purposes, and evaluated them in a quantitative manner:

1. Which are the five top-ranked terms that occur on the web pages mentioning "Iron Maiden"?
2. Indicate the number of web pages containing all of the terms "Iron Maiden", "metal", and "guitar".
3. Show a list of web pages that contain the terms "Iron Maiden" and "british".
4. Considering the complete set of web pages, which are the three terms that co-occur on the highest number of web pages?
5. How many web pages contain the terms "Iron Maiden" and "metal", but not the term "guitar"?
6. Display a list of audio files available at web pages containing the term "Iron Maiden".
7. Which terms co-occur on the set of web pages that contains the highest number of image files in hierarchy level three?

8. Indicate the URL of one particular web page that contains image files but no video files.
9. How many web pages does the complete collection contain?
10. Find one of the deepest elements in the hierarchy and select it.
11. Generate a new visualization using only the web pages on which the terms "bass" and "heavy metal" co-occur.

The tasks 1–8 are general ones that are likely to arise when analyzing and browsing collections of web pages. In particular, tasks 1–5 address the co-occurring terms, whereas tasks 6–8 deal with the multimedia content extracted from the web pages. In contrast, the tasks 9–11 relate to the structure of the Sunburst tree.

After having explained the interaction functionalities provided by 3D-COB to our participants, they had five minutes to explore the user interface themselves with a visualization gained for "Britney Spears". During this warm-up, the participants were allowed to ask questions. After the exploration phase, we presented them the visualization obtained when using the web page collection of "Iron Maiden", cf. Figure 1. We consecutively asked them each of the questions and measured the time they needed to finish each task. Each participant had a maximum time of three minutes to complete each task. The constraints were set as follows: $N = 8$, $R = 8$, and $E = 3.0$ (cf. Subsection 3.2).

Due to time limitations, we had to restrict the number of participants in the user study to six (five males, one female). All of them were computer science or business students at [omitted due to blind review] and all stated to have a moderate or good knowledge of user interfaces and to be very interested in music. All participants performed the user study individually, one after another.

5.2 Results and Discussion

As for the results of the study, Table 2 shows the time, in seconds, needed by the participants (A–F) to finish each task. In general, the tasks related to structural questions were answered in a shorter time than those related to browsing the collection. Among the structural questions, solely task 11 required a quite high average time. This is explained by the fact that the term "bass" was not easy to find on all layers. The same holds for the term "british" requested in task 3.

For the questions related to browsing in the hierarchy, it was observed that tasks requiring intensive rotation of the Sunburst stack (1, 3, 4, 5, 7) yielded worse results than those for which this was not the case (2, 6). In general, users spent a lot of time rotating the Sunburst stack to a position at which the label of the selected arc was readable. This process should be automatized in future versions of 3D-COB.

The relatively high average time required to perform the first task may be attributed to the fact that most participants needed some seconds to get used to the new visualization of "Iron Maiden" after having explored the web pages of "Britney Spears" in the exploration phase. In spite of the fact that task 3 was solved in only 37 seconds on average, we realized that some participants

Table 2. For each participant, the time (in seconds) needed to finish each task of the user study on ergonomic aspects. Inverse numbers indicate that the given answer was wrong. The mean was calculated excluding the wrong answers.

Task	1	2	3	4	5	6	7	8	9	10	11
A	28	13	45	47	36	61	172	*180*	2	12	*25*
B	69	23	46	52	14	15	68	76	*6*	12	62
C	15	3	39	27	22	3	34	68	1	9	31
D	132	1	57	30	117	14	43	*180*	5	12	40
E	110	9	16	8	*163*	7	12	148	2	38	74
F	36	14	21	46	44	12	79	*180*	3	5	61
Mean	65	11	37	35	47	19	68	97	3	15	54

had problems locating the arc "british" since it was hardly perceivable due to its position behind a much higher arc. As both task 4 and 2 required finding the same arc, it was quite interesting that the averaged times differed considerably. As for task 5, two participants were not sure which number to subtract from which other. Except for one participant, who chose a correct but time-consuming solution, task 6 was generally solved quickly. Solving task 7 took the second highest average time since it required finding and navigating to the Sunburst that illustrates the amount of image files and comparing the heights of all arcs in hierarchy level three of this Sunburst. Task 8 yielded the worst results as no arc on the video layer had a height of zero, which confused most of the participants. It was obviously not clear that a positive height of an arc on the video layer does not necessarily mean that each web page represented by this arc offers video content.

To conclude, the user study assessing ergonomic aspects showed that 3D-COB can be efficiently used for tasks related to browsing sets of web pages. Although barely comparable to the user study on similar tree visualization systems conducted in [14], due to a different application scenario, a very rough comparison of the average total performance times for the tasks shows that this time is much shorter for 3D-COB (45 sec) than for the best performing system of [14] (101 sec). Therefore, our results are promising. However, it can be stated that the user interaction functionalities provided by 3D-COB need some improvements. This will be addressed in future versions of the implementation.

6   Conclusions and Future Work

In this paper, we presented the *Three-Dimensional Co-Occurrence Browser (3D-COB)*, a user interface for browsing collections of music artist-related web pages in a novel way. 3D-COB automatically extracts musically relevant terms from web pages about artists, applies a term weighting function, organizes the web pages according to co-occurring terms, and finally employs a variant of the Sunburst visualization technique to illustrate not only the extracted terms, but also the amount of multimedia files, grouped in different categories (audio, image, video).

Moreover, we conducted two user studies: one to evaluate the performance of different term weighting strategies for finding descriptive artist terms, the second to assess ergonomic aspects of 3D-COB's user interface. From the first study, we learned that using TF·IDF yielded significantly worse results than the simple TF and DF measures with respect to the appropriateness of the terms to describe the music artists used in our experiments. In contrast, comparing the measures TF and DF, no significant difference in their performance was detected. The second user study showed that 3D-COB offers valuable additional information about web pages that cannot be discovered by the standard list-based representation of search results, which is commonly used by web search engines.

As for future work, we will elaborate alternative ways to navigate in the visualization, e.g., using alternative input devices. We are also developing a focused web crawler in combination of which 3D-COB may be used to browse a set of web pages related to a certain topic (not necessarily related to music) without relying on existing search engines. The 3D-COB could also be employed to domains like movies, literature, painting, or news. Moreover, we will improve the user feedback provided by 3D-COB, e.g., by showing all terms that co-occur in the selected set of web pages, independently from the labels of the arcs. Finally, smoothly embedding the multimedia content directly in the user interface instead of opening it in external applications would be a desirable feature for future versions of 3D-COB.

Acknowledgments

This research is supported by the Austrian Fonds zur Förderung der Wissenschaftlichen Forschung (FWF) under project number L511-N15.

References

1. http://dir.yahoo.com/Entertainment/Music/Genres (March 2008)
2. http://en.wikipedia.org/wiki/List_of_file_formats (December 2007)
3. http://finetunes.musiclens.de (March 2008)
4. http://last.fm (December 2007)
5. http://www.allmusic.com (November 2007)
6. http://www.processing.org (March 2008)
7. http://www.wikipedia.org (December 2007)
8. Andrews, K., Heidegger, H.: Information Slices: Visualising and Exploring Large Hierarchies using Cascading, Semi-Circular Discs. In: Proc. of IEEE Information Visualization 1998, Research Triangle Park, NC, USA (1998)
9. Geleijnse, G., Korst, J.: Web-based Artist Categorization. In: Proc. of the 7th Int'l Conf. on Music Information Retrieval, Victoria, Canada (2006)
10. Goto, M., Goto, T.: Musicream: New Music Playback Interface for Streaming, Sticking, Sorting, and Recalling Musical Pieces. In: Proc. of the 6th Int'l Conf. on Music Information Retrieval, London, UK (2005)
11. Johnson, B., Shneiderman, B.: Tree-Maps: A Space-Filling Approach to the Visualization of Hierarchical Information Structures. In: Proc. of the 2nd IEEE Conf. on Visualization, San Diego, CA, USA (1991)

12. Knees, P., Pampalk, E., Widmer, G.: Artist Classification with Web-based Data. In: Proc. of the 5th Int'l Symposium on Music Information Retrieval, Barcelona, Spain (2004)
13. Knees, P., Schedl, M., Pohle, T., Widmer, G.: An Innovative Three-Dimensional User Interface for Exploring Music Collections Enriched with Meta-Information from the Web. In: Proc. of the 14th ACM Conf. on Multimedia 2006, Santa Barbara, CA, USA (2006)
14. Kobsa, A.: User Experiments with Tree Visualization Systems. In: Proc. of the 10th IEEE Symposium on Information Visualization 2004, Austin, Texas, USA (2004)
15. Mörchen, F., Ultsch, A., Nöcker, M., Stamm, C.: Databionic Visualization of Music Collections According to Perceptual Distance. In: Proc. of the 6th Int'l Conf. on Music Information Retrieval, London, UK (2005)
16. Pampalk, E., Flexer, A., Widmer, G.: Hierarchical Organization and Description of Music Collections at the Artist Level. In: Proc. of the 9th European Conf. on Research and Advanced Technology for Digital Libraries, Vienna, Austria (2005)
17. Pampalk, E., Goto, M.: MusicSun: A New Approach to Artist Recommendation. In: Proc. of the 8th Int'l Conf. on Music Information Retrieval, Vienna, Austria (2007)
18. Schedl, M., Knees, P., Seyerlehner, K., Pohle, T.: The CoMIRVA Toolkit for Visualizing Music-Related Data. In: Proc. of the 9th Eurographics/IEEE VGTC Symposium on Visualization, Norrköping, Sweden (2007)
19. Schedl, M., Pohle, T., Knees, P., Widmer, G.: Assigning and Visualizing Music Genres by Web-based Co-Occurrence Analysis. In: Proc. of the 7th Int'l Conf. on Music Information Retrieval, Victoria, Canada (2006)
20. Sheskin, D.J.: Handbook of Parametric and Nonparametric Statistical Procedures, 3rd edn. Chapman & Hall/CRC, Boca Raton (2004)
21. Stasko, J., Zhang, E.: Focus+Context Display and Navigation Techniques for Enhancing Radial, Space-Filling Hierarchy Visualizations. In: Proc. of IEEE Information Visualization 2000, Salt Lake City, UT, USA (2000)
22. Vignoli, F., van Gulik, R., van de Wetering, H.: Mapping Music in the Palm of Your Hand, Explore and Discover Your Collection. In: Proc. of the 5th Int'l Symposium on Music Information Retrieval, Barcelona, Spain (2004)
23. Whitman, B., Lawrence, S.: Inferring Descriptions and Similarity for Music from Community Metadata. In: Proc. of the 2002 Int'l Computer Music Conference, Göteborg, Sweden (2002)
24. Zadel, M., Fujinaga, I.: Web Services for Music Information Retrieval. In: Proc. of the 5th Int'l Symposium on Music Information Retrieval, Barcelona, Spain (2004)

Semantic-Driven Multimedia Retrieval with the MPEG Query Format

Ruben Tous and Jaime Delgado

Distributed Multimedia Applications Group (DMAG)
Universitat Politecnica de Catalunya (UPC), Dpt. d'Arquitectura de Computadors
rtous@ac.upc.edu, jaime.delgado@ac.upc.edu

Abstract. The MPEG Query Format (MPQF) is a new standard from
the MPEG standardization committee which provides a standardized interface to multimedia document repositories. The purpose of this paper is
describing the necessary modifications which will allow MPQF to manage
metadata modelled with Semantic Web languages like RDF and OWL,
and query constructs based on SPARQL. The suggested modifications
include the definition of a new MPQF query type, and a generalization
of the MPQF metadata processing model. As far as we know, this is the
first work to apply the MPEG Query Format to semantic-driven search
and retrieval of multimedia contents.

1 Introduction

The research around multimedia information retrieval (MIR) is gaining relevance
due to the increasing amount of digitally stored multimedia contents and the
evolution of the ways of managing and consuming them. The main goals are to
enable efficient, precise and expressive ways to query standalone and distributed
multimedia data collections. This responds to the necessity to fullfill the new
requirements imposed by the scaling of traditional collections, but specially to
satisfy new requirements related to novel usages.

1.1 Multimedia Search and Retrieval: Information Retrieval and Data Retrieval

The MIR process usually starts with an end-user expressing his information
needs through a human-friendly query interface. User's information needs come
from the conceptual level, and combine criteria about the information represented by the content's data, i.e. the *meaning* of the content (e.g. *"images of
a bird"*) with other criteria about the features of the content's data itself (e.g.
"image files of less than one megabyte"). At the end, the only way of fulfilling
the user information needs is translating these criteria into machine-readable
conditions, as precise as possible, over the content's data, being these data the
media binary representation or some metadata annotations. Metadata annotations provide information about the content at different levels, from low-level
features and management information, to semantic-level descriptions.

D. Duke et al. (Eds.): SAMT 2008, LNCS 5392, pp. 149–163, 2008.

So, the problem is twofold, in one hand we have the challenge of enriching the media data with metadata useful for solving the queries. On the other hand we need to face the problem of formalizing the user information needs, as far as possible, in the form of a query expressed in terms of the available metadata model. Even in we succeed in the first challenge, the second remains non trivial, because end-users expresses their criteria performing actions over the provided human-friendly user interface. While some criteria can be easy to formalize, *semantic-level* criteria use to be more difficult to express and also to be solved in a deterministic way. For these situations, formal conditions are usually combined with non-formal or "fuzzy" query terms (e.g. Query By Example, Query By Keywords) and Information Retrieval techniques are applied.

So, MIR systems have the particularity that they must combine Information Retrieval (IR) techniques, with techniques for querying metadata, which belong to the Data Retrieval (DR) area within the Databases discipline. Both approaches aim to facilitate users access to information, but from different points-of-view. On one hand, an Information Retrieval system aims to retrieve information that might be relevant to the user even though the query is not formalized, or the criteria are fuzzy. In contrast, a Data Retrieval system (e.g. an XQuery-based database) deals with a well defined data model and aims to determine which objects of the collection satisfy clearly defined conditions (e.g. the title of a movie, the size of a video file or the fundamental frequency of an audio signal).

1.2   MPEG Query Format (MPQF)

Recently, the MPEG standardization committee (ISO/IEC JTC1/SC29/WG11) is developing a new standard, the MPEG Query Format (MPQF) [1,4], which aims to provide a standardized interface to multimedia document repositories. MPQF defines the format of queries and responses between parties in a multimedia search and retrieval process. In one hand, standardizing such kind of language fosters *interoperability* between parties in the multimedia value chain (e.g. content providers, aggregators and user agents). On the other hand, MPQF favours also *platform independence*; developers can write their applications involving multimedia queries independently of the system used, which fosters software reusability and maintainability.

One of the key features of MPQF is that it is designed for expressing queries combining the expressive style of Information Retrieval (IR) systems (e.g. query-by-example and query-by-keywords) with the expressive style of XML Data Retrieval (DR) systems (e.g. XQuery [17]), embracing a broad range of ways of expressing user information needs. Regarding IR-like criteria, MPQF offers a broad range of possibilities that include but are not limited to *QueryByDescription* (query by example description), *QueryByFreeText*, *QueryByMedia* (query by example media), *QueryByROI* (query by example region of interest), *QueryByFeatureRange*, *QueryBySpatialRelationships*, *QueryByTemporalRelationships* and *QueryByRelevanceFeedback*. Regarding DR-like criteria, MPQF offers its own XML query algebra for expressing conditions over the multimedia related

XML metadata (e.g. Dublin Core , MPEG-7 or any other XML-based metadata format) but also offers the possibility to embed XQuery expressions.

1.3   Semantic-Driven Multimedia Information Retrieval

So, a MIR system will deal with two related but different challenges, IR and DR. Though there is a solid research basis regarding the Information Retrieval challenge, the necessity to face such problem appears, in fact, because of the difficulty of annotating the content with the necessary metadata and the difficulty of formalizing the end-user's *semantic-level* criteria. The gap between low-level content description and querying, and the related high-level or semantic description and querying is known as the *semantic gap*. As a result, from the multimedia retrieval point-of-view, measures are needed to deal with uncertainty and the potential lack of search precision. However, in a vast number of scenarios, simple IR-like mechanisms like keywords-based search use to offer pretty satisfactory results even when the size of the target collections is big.

There are, nevertheless, situations in which the end-user requirements, and/or the circumstances, motivate the efforts of producing higher-level metadata descriptors (semantic or not) and formalizing parts of the user's *semantic-level* criteria moving them to the Data Retrieval realm. An example could be the video surveillance scenario, in which a huge quantity of information is stored, and the query expressiveness and results precision are critical. This *formalization* task requires enhancing the metadata production layer but also implies offering to the user a richer interface or, in subsequent layers, post-processing the initial non-formalized query. This enrichment of the querying process is related to the improvement of the metadata-level query capabilities. The result is the starting point of what is known as *semantic-driven* Multimedia Information Retrieval, whose evolution leads to the usage of *semantic-specific* technologies as those from the Semantic Web initiative.

1.4   MPQF Limitations Managing Semantic Web Related Metadata

MPQF is an XML-based language in the sense that all MPQF instances (queries and responses) must be XML documents. Formally, MPQF is Part 12 of ISO/IEC 15938, "Information Technology - Multimedia Content Description Interface" better known as MPEG-7 [8]. However, the query format was technically decoupled from MPEG-7 and is now metadata-neutral. So, MPQF is not coupled to any particular metadata standard, but it assumes that any metadata related to multimedia content being modelled as an XML Infoset. So, this metadata *neutrality* is constrained by the fact that MPQF expresses conditions and projections over the metadata using XPath expressions, i.e. privileging XML-enabled metadata repositories but restraining those based in other models, specially those based in RDF [10] metadata.

In this paper we describe the necessary modifications which will allow MPQF to manage metadata modelled with Semantic Web languages like RDF and OWL [9], and query constructs based on SPARQL [14]. The suggested modifications include the definition of a new MPQF query type, and a generalization of the MPQF metadata processing model.

2   MPQF in Depth

2.1   MPEG Query Format Syntax and Terminology

Before entering the discussion of how the MPQF syntax and data model could be adapted for semantic-driven retrieval, let's briefly describe the structure of MPQF instances. MPQF queries (requests and responses) are XML documents that can be validated against the MPQF XML schema (see Figure 1). MPQF instances include always the *MpegQuery* element as the root element. Below the root element, an MPQF instance includes the *Query* element or the *Management* element. MPQF instances with the *Query* element are the usual requests and responses of a digital content search process. The *Query* element can include the *Input* element or the *Output* element, depending if the document is a request or a response. The part of the language describing the contents of the *Input* element (requests) is named the Input Query Format (IQF) in the standard. The part of the language describing the *Output* element (responses) is named the Output Query Format (OQF) in the standard. IQF and OQF are just used to facilitate understanding, but do not have representation in the schema. Alternatively, below the root element, an MPQF document can include the *Management* element. Management messages (which in turn can be requests and responses) provide means for requesting service-level functionalities like interrogating the capabilities of a MPQF processor.

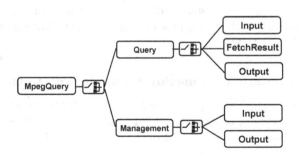

Fig. 1. MPQF language parts

Example in Code 1 shows an input MPQF query asking for JPEG images related to the keyword *"Barcelona"*.

2.2   MPEG Query Format Database Model

As happens with any other query language, an MPQF query is expressed in terms of a specific database model. The MPQF database model formally defines

Code 1. Example MPQF input query

```
<MpegQuery>
  <Query>
    <Input>
      <OutputDescription outputNameSpace="//purl.org/dc/elements/1.1/">
        <ReqField>title</ReqField>
        <ReqField>date</ReqField>
      </OutputDescription>
      <QueryCondition>
        <TargetMediaType>image/jpg</TargetMediaType>
        <Condition xsi:type="AND" preferenceValue="10">
          <Condition xsi:type="QueryByFreeText">
            <FreeText>Barcelona</FreeText>
          </Condition>
          <Condition xsi:type="GreaterThanEqual">
            <DateTimeField>date</DateTimeField>
            <DateValue>2008-01-15</DateValue>
          </Condition>
        </Condition>
      </QueryCondition>
    </Input>
  </Query>
</MpegQuery>
```

the information representation space which constitutes the evaluation basis of
an MPQF query processor. MPQF queries are evaluated against one or more
multimedia databases which, from the point-of-view of MPQF, are unordered
sets of *Multimedia Contents*. The concept of *Multimedia Content* (MC) refers
to the combination of multimedia data (resource) and its associated metadata.
MPQF allows retrieving complete or partial MC's data and metadata by specification of a condition tree. So, MPQF deals with a dual database model (see
Figure 2) constituted by content and metadata.

Example in Figure 3 shows a graphical representation of an MPQF condition
tree carrying two different conditions which reflect the duality of the database
model used by the language. In one hand, there's an IR-like condition using the

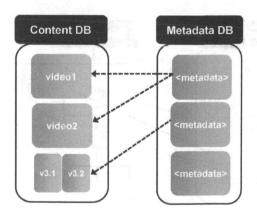

Fig. 2. Dual database model

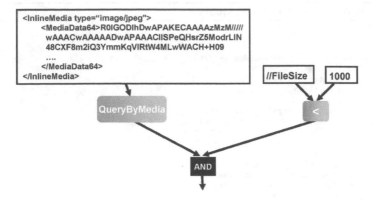

```
<InlineMedia type="image/jpeg">
    <MediaData64>R0lGODlhDwAPAKECAAAAzMzM/////
    wAAACwAAAAADwAPAAACIISPeQHsrZ5ModrLIN
    48CXF8m2iQ3YmmKqVlRtW4MLwWACH+H09
    ....
    </MediaData64>
</InlineMedia>
```

Fig. 3. Example condition tree

QueryByExample query type and including the Base64 encoding of the binary contents of an example JPEG image. On the other hand, there is a simple DR-like condition specifying that the metadata field *FileSize* must be inferior to 1000 bytes.

In order to deal this model duality, MPQF operates over sequences of what the standard calls *evaluation-items*. By default, an *evaluation-item* (EI) is a multimedia content in the multimedia database, but other types of EIs are also possible. For instance, an EI can be a segment of a multimedia resource, or an XPath-item related to a metadata XML tree. The scope of query evaluation and the granularity of the result set (the granularity of EIs) can be determined by a *EvaluationPath* element specified within the query. If this *EvaluationPath* element is not specified, the output result is provided as a collection of multimedia contents, as stored in the repository, all satisfying the query condition.

The *EvaluationPath* element determines the query scope on basis to a hypothetical XML metadata tree covering the entire database. So, we can redrawn

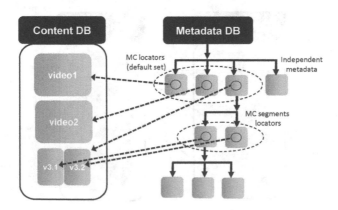

Fig. 4. Hierarchical MPQF model

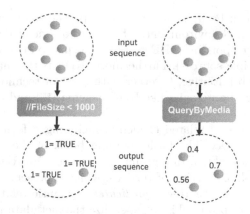

Fig. 5. Two evaluation styles in MPQF: Boolean and fuzzy-logic

our visual representation of the model in order to show this hierarchical nature, as shown in Figure 4.

2.3   MPQF Evaluation Model

The condition tree of an MPQF query is constructed combining filtering elements (conditions) from the *BooleanExpressionType* and interconnecting them with Boolean operators (*AND, OR, NOT* and *XOR*). Each condition acts over a sequence of evaluation-items and, for each one, return a value in the range of [0..1]. In the case of DR-like conditions (e.g. it "the size of the file must be smaller than 1000 bytes") the condition can return just 1 or 0, which mean *true* and *false* respectively. In the case of IR-like conditions, they can evaluate to any value in the range of [0..1]. A *threshold* value within a condition is used to indicate the minimum value the score of an evaluation-item is required to have. Otherwise the evaluation-item is not considered further during evaluation.

So, with respect to DR-like conditions, MPQF acts as a conventional Boolean-based filtering language, while with respect to IR-like conditions MPQF acts preserving scores as a fuzzy-logic system. The standard specifies the behaviour of the provided Boolean operators in presence of non-Boolean values.

3   An MPQF Extension to Manage Semantic-Modelled Metadata

3.1   Metadata and RDF

Current practices in the metadata community show an increasing usage of Semantic Web technologies like RDF and OWL. Some relevant initiatives are choosing the RDF language (e.g. [2]) for modelling metadata (semantic or not) because of its advantages with respect to other formalisms. RDF is modular; a subset of RDF triples from an RDF graph can be used separately, keeping a consistent

RDF model. So it can be used in presence of partial information, an essential feature in a distributed environment. The union of knowledge is mapped into the union of the corresponding RDF graphs (information can be gathered incrementally from multiple sources). Furthermore, RDF is the main building block of the Semantic Web initiative, together with a set of technologies for defining RDF vocabularies and ontologies like RDF Schema [12] and the Ontology Web Language (OWL) [9].

RDF comprises several related elements, including a formal model, and an XML serialization syntax. The basic building block of the RDF model is the triple *subject-predicate-object*. In a graph theory sense, an RDF instance is a labelled directed graph, consisting of vertices, which represent *subjects* or *objects*, and labelled edges, which represent *predicates* (semantic relations between *subjects* and *objects*). Our proposal is to generalize the metadata part of the MPQF database model to allow the presence (in requests and responses) of RDF predicates. Example in Code 2 shows metadata including semantic information about a JPEG image in the triples notation [10].

Code 2. Example RDF graph using the triples notation

```
example:bigbird.jpg    dc:creator                "Ruben Tous .
example:bigbird.jpg    example:terms/represents  example:terms/Eagle .
example:terms/Eagle    rdfs:subClassOf           example:terms/Bird .
```

3.2   OWL, Ontologies, and Knowledge Bases

OWL is a vocabulary for describing properties and classes of RDF resources, complementing the RDF Schema (RDFS) capabilities in providing semantics for generalization-hierarchies of such properties and classes. OWL enriches the RDFS vocabulary by adding, among others, relations between classes (e.g. disjointness), cardinality (e.g. "exactly one"), equality, richer typing of properties, characteristics of properties (e.g. symmetry), and enumerated classes. OWL has the influence of more than 10 years of Description Logic research, and is based on the SH family of Description Logics [5]. The language has three increasingly expressive sublanguages designed for different uses: OWL Lite, OWL DL and OWL Full.

The combination of OWL ontologies with RDF statements compliant with the vocabularies and constraints defined in these ontologies make up what is known as a knowledge base (KB). The terms *Tbox* (terminological component) and *Abox* (assertion component) are used to describe these two different types of statements in a KB (see Figure 6).

3.3   SPARQL

SPARQL [14] is a popular RDF query language which consists of triple patterns. SPARQL is based in RDQL [13] from HP Labs Bristol, which in turn was based

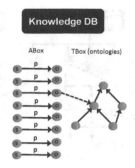

Fig. 6. Knowledge Base (KB)

in SquishQL [15], derived from rdfDB [11]. SPAQL is standardized by the RDF Data Access Working Group (DAWG) of the World Wide Web Consortium, and has become an official W3C Recommendation on 15th January 2008.

An SPARQL query consists of a *basic graph pattern*, expressed as a list of triple patterns. Triple patterns are like RDF triples except that each of the subject, predicate and object may be a variable. Each triple pattern is comprised of named variables and RDF values (URIs and literals). An SAPQRL query can additionally have a set of constraints on the values of those variables, and a list of the variables required in the answer set. An example SPARQL query is shown in Code 3.

Code 3. Example SPARQL query

```
PREFIX   dc:   <http://purl.org/dc/elements/1.1/>
PREFIX   ns:   <http://example.org/ns#>
SELECT   ?title ?price
WHERE    { ?x ns:price ?price .
           FILTER (?price < 30.5)
           ?x dc:title ?title . }
```

Each solution to the example query will expose one way in which the selected variables can be bound to RDF terms so that the query pattern matches the data. The result set gives all the possible solutions. We envisage that SPARQL will become the standard language for querying semantic metadata. In the following sections we describe the necessary actions to allow SPARQL expressions to be embedded within MPQF queries, in a similar way that XQuery expressions can.

3.4   MPQF Database Model and Semantic-Modelled Metadata

We defend that the MPQF model should be generalized in such a way that semantic metadata, modelled with knowledge representation techniques such as RDF and OWL, could be also used in query conditions and resultsets. MPQF strongly relies in the existence (real or just logical) of an XML metadata hierarchy acting as a catalogue for all the multimedia contents and their related

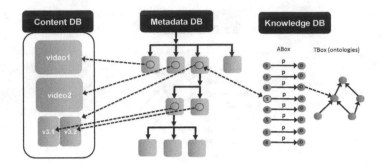

Fig. 7. MPQF model and semantic metadata

segments. Of course our approach does not try modifying this feature, as it is an essential part of MPQF, and our aim is just extending the language. A certain multimedia database could combine an XML metadata catalogue with a knowledge base composed by a *Tbox* and an *Abox*. Both kinds of metadata (XML catalogue and the KB) could be managed separately, and the binding between metadata descriptions could be the URIs of the media locators (see Figure 7).

3.5   A New Query Type: QueryBySPARQL

Complex conditions in MPQF are encapsulated within elements from the *Query-Type*. This type is an abstract type which extends the *BooleanExpressionType* and is the parent class of the *QueryByMedia*, *QueryByDescription*, *QueryByFeatureRange*, *SpatialQuery*, *TemporalQuery*, *QueryByXQuery*, *QueryByFreeText*, and *QueryByRelevanceFeedback* types.

We suggest incorporating a new query type, the *QueryBySPARQL*, which will inherit from the *QueryType* and will act in a similar way as *QueryByXQuery* type. We suggest adding the declaration appearing in Code 4 within the MPQF XML schema.

Code 4. New *QueryBySPARQL* type declaration

```
<complexType name="QueryBySPARQL">
  <complexContent>
    <extension base="mpqf:QueryType">
      <sequence>
        <element name="SPARQLQuery" type="string"/>
      </sequence>
    </extension>
  </complexContent>
</complexType>
```

Table 1 includes the proposed normative definition of the semantics of all the components of the new query type specification.

Code 5 shows an example MPQF input query with an SPARQL query embedded in it. The embedded query makes use of the *ASK* clause described in [14].

Table 1. Semantics of the new *QueryBySPARQL* type

Name	Definition
QueryBySPARQL	Extends the abstract *QueryType* type and denotes a query operation for the use of SPARQL expressions
SPARQLQuery	Specifies the SPARQL expression that is used to filter information. Because this query type only can return *true* or *false*, not all possible SPARQL espressions are allowed (no output generation is allowed in query types). The only valid SPARQL expressions te be used using the SPARQL *ASK* clause and the *?resource* variable appearing unbounded within the graph pattern. The variable *?resource* will be the reference to the evaluation item being processed. If the query pattern matches the RDF data stored in the database (or can be inferred from it) the result will be *true*. Otherwise it will be *false*.

This clause serves to test whether or not a query pattern has a solution. No information is returned about the possible query solutions, just whether or not a solution exists. Figure 8 shows graphically the behaviour of new *QueryBySPARQL* query type. The evaluation-item being processed, the one highlighted within the metadata XML tree, evaluates to *true* because there is a graph of triples matching the given pattern.

Code 5. Example query with a *QueryBySPARQL* condition

```
<MpegQuery mpqfID="someID">
  <Query>
    <Input>
      <QueryCondition>
        <Condition xsi:type="QueryBySPARQL">
          <XQuery>
                  <![CDATA[
                  PREFIX  dc:  <http://purl.org/dc/elements/1.1/>
                  ASK     { ?resource dc:title "Barcelona . }
                  ]]>
          </XQuery>
        </Condition>
      </QueryCondition>
    </Input>
  </Query>
</MpegQuery>
```

3.6 RDF Triples in the Resultset

The ability to express SPARQL expressions within MPQF queries would not be very useful if one could not know which RDF triples are related to an item. Because the allowed SPARQL expressions are constrained to return *true* or *false*, another mechanism is required in order to allow users to query the RDF data in the database. Of course this could be done through an independent RDF query interface, but we believe that it could be interesting if MPQF queries could return also RDF metadata.

The suggested extension implies two different aspects, the MPQF Input Query Format and the MPQF Output Query Format. In one hand, the current MPQF

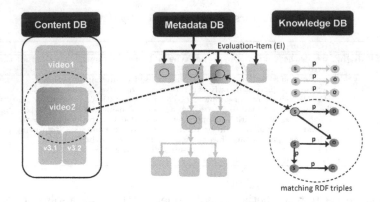

Fig. 8. Visual representation of the execution of the new *QueryBySPARQL* query type

Input Query Format allows specifying which is the desired data to be returned within the result records. This is the part of the Input Query Format called *output description*. Code 6 shows an example MPQF input query with its corresponding output description.

Code 6. Output description example

```
<MpegQuery mpqfID="someID">
 <Query>
  <Input>
   <OutputDescription freeTextUse="true" outputNameSpace="urn:mpeg:mpeg7:schema:2004" >
    <ReqField typeName="CreationInformationType">/Creation/Title</ReqField>
    <ReqField typeName="CreationInformationType">/Creation/Creator</ReqField>
    <ReqField typeName="MediaFormatType">/FileFormat</ReqField>
   </OutputDescription>
   <QueryCondition>
    <Condition xsi:type="QueryByFreeText">
     <FreeText>San Jose</FreeText>
    </Condition>
   </QueryCondition>
  </Input>
 </Query>
</MpegQuery>
```

Currently there are two different mechanisms for describing the output, some fixed attributes (*freeTextUse*, *mediaResourceUse*, *thumbnailUse* ,etc.) and user defined *ReqField* elements which carry an XPath expression. So, there are two possible solutions in order to allow describing RDF output. The simple one could consist on just adding a *RDFUse* optional attribute which would signal the database to return all RDF triples related to the resource. However, there could be also a more flexible solution which would consist on extending the semantics of the *ReqField* element in order to allow selective harvesting of RDF metadata.

On the other hand, it is also necessary to consider where the resulting RDF triples would be placed within an output query. Fortunately, the MPQF Output Query Format is very flexible in that sense, and it allows the inclusion of XML metadata from any format. Making use of the available XML serialization for RDF, we can embed the RDF triples within an MPQF resultset, as shown in Code 7.

Code 7. Example showing RDF data within an output MPQF query

```
<MpegQuery mpqfID="someID">
  <Query>
    <Output currPage="1" totalPages="1" expirationDate="2008-05-30T09:00:00">
      <ResultItem xsi:type="ResultItemType" recordNumber="1">
        <TextResult>Item 01</TextResult>
        <MediaResource>http://dmag.upf.es/mpqf/repository/item01.avi</MediaResource>
        <Description xmlns:dc="http://purl.org/dc/elements/1.1/"/>
          <dc:title>Barcelona</dc:title>
          <dc:creator>John Smith</dc:creator>
          <dc:format>image/jpeg</dc:format>
        </Description>
      </ResultItem>
    </Output>
  </Query>
</MpegQuery>
```

---

4 Related Work

As far as we know, this is the first work to apply the MPEG Query Format to semantic-driven search and retrieval of multimedia contents. The initial work related to MPQF can be found in 2004, as a cooperation between MPEG (ISO/IEC JTC 1/SC 29 WG11), its MPEG-7 standard and JPEG (ISO/IEC JTC 1/SC 29 WG1) and its JPSearch initiative. In July 2006, during the 77th MPEG meeting, the MPEG committee released a final Call for Proposal and specified a final set of requirements. MPQF reached the status of Final Draft International Standard (FDIS) after the 84th MPEG meeting, April 2008. It is expected that MPQF will become an ISO standard after the 85th MPEG meeting, July 2008.

Regarding other multimedia query formats, there exist several old languages explicitly for multimedia data such as SQL/MM [7], which have limitations in handling XML data and video contents (generally because of the date in which their design started). In the last years, these kind of languages have been usually based on MPEG-7 descriptors and the MPEG-7 data model. Some simply defend the use of XQuery or some extensions of it. Others define a more high-level and user-oriented approach. MPQF outperforms XQuery-based approaches like [6,16,3] because, while offering the same level of expressiveness, it offers multiple content-based search functionalities (QBE, query-by-freetext) and other IR-like features (e.g. paging or relevance feedback). Besides, XQuery does not provide means for querying multiple databases in one request and does not support multimodal or spatial/temporal queries.

Another novel feature of MPQF is its *metadata-neutrality*, which allows using it over metadata formats other than MPEG-7. However, as pointed in previous sections, this *metadata-neutrality* is constrained by the fact that MPQF is currently limited to be used with XML-based metadata formats. Our work describe the necessary modifications which will allow MPQF to manage metadata modelled with Semantic Web languages like RDF and OWL, and query constructs based on SPARQL.

5   Conclusions and Future Work

The MPEG Query Format (MPQF) is one of the new standardization initiatives from the MPEG standardization committee, which aims providing a standardized interface to multimedia document repositories. In this paper we have described the necessary extensions which will allow MPQF to manage metadata modelled with Semantic Web languages like RDF and OWL, and query constructs based on SPARQL, an RDF query language. The described extensions include a conceptual generalization of the MPQF metadata processing model, the definition of a new MPQF query type (*QueryBySPARQL*), and a minor extension of the *output description* part of the MPQF's Input Query Format.

Currently there is an increasing interest in the advantages of having multimedia contents annotated with Semantic Web languages and bounded to multimedia ontologies. This evolution strongly impacts the way multimedia contents are searched and retrieved in a broad range of application scenarios. Our approach opens the possibility to use a modern standard multimedia query language like MPQF in combination with these new semantic metadata modelling technologies.

Acknowledgements

This work has been partly supported by the Spanish government (DRM-MM project, TSI 2005-05277) and the European Network of Excellence VISNET-II (IST-2005-2.41.6), funded under the European Commission IST 6th Framework Program.

References

1. Adistambha, K., Doeller, M., Tous, R., Gruhne, M., Sano, M., Tsinaraki, C., Christodoulakis, S., Yoon, K., Ritz, C., Burnett, I.: The MPEG-7 Query Format: A New Standard in Progress for Multimedia Query by Content. In: Proceedings of the 7th International IEEE Symposium on Communications and Information Technologies (ISCIT 2007), pp. 479–484, Sydney, Australia (2007)
2. The dublin core metadata initiative (dcmi), http://hublincore.org/
3. Fatemi, N., Khaled, O.A., Coray, G.: An xquery adaptation for mpeg-7 documents retrieval, http://www.idealliance.org/papers/dx_xml03/papers/ 05-03-0105-03-01.html

4. Gruhne, M., Tous, R., Döller, M., Delgado, J., Kosch, H.: MP7QF: An MPEG-7 Query Format. In: Proceedings of the 3rd International Conference on Automated Production of Cross Media Content for Multi-channel Distribution (AXMEDIS 2007), Barcelona, Spain, pp. 15–18 (2007)
5. Horrocks, I., Sattler, U., Tobies, S.: Practical reasoning for very expressive description logics. J. of the Interest Group in Pure and Applied Logic (2000)
6. Kang, J., et al.: An xquery engine for digital library systems. In: 3rd ACM/IEEE-CS Joint Conference on Digital Libraries, Houston, Texas (May 2003)
7. Melton, J., Eisenberg, A.: SQL Multimedia Application packages (SQL/MM). ACM SIGMOD Record 30(4), 97–102 (2001)
8. Iso/iec 15938 version 2. information technology - multimedia content description interface (mpeg-7) (2004)
9. Owl web ontology language overview. w3c recommendation (February 10, 2004), http://www.w3.org/TR/owl-features/
10. Resource description framework, http://www.w3.org/RDF/
11. rdfdb query language, http://www.guha.com/rdfdb/query.html
12. Rdf vocabulary description language 1.0: Rdf schema, http://www.w3.org/TR/2003/WD-rdf-schema-20030123/
13. Rdql - a query language for rdf. w3c member submission (January 9, 2004), http://www.w3.org/Submission/RDQL/
14. Sparql query language for rdf. w3c working draft (October 12, 2004), http://www.w3.org/TR/rdf-sparql-query/
15. Inkling: Rdf query using squishql, http://swordfish.rdfweb.org/rdfquery/
16. Tjondronegoro, D., Chen, Y.P.P.: Content-based indexing and retrieval using mpeg-7 and xquery in video data management systems. World Wide Web: Internet and Web Information Systems, pp. 207–227 (2002)
17. Xquery 1.0: An xml query language. w3c proposed recommendation (November 21, 2006), http://www.w3.org/TR/xquery/

On Video Abstraction Systems' Architectures and Modelling

Víctor Valdés and José M. Martínez

Video Processing and Understanding Lab
Escuela Politécnica Superior, Universidad Autónoma de Madrid, Spain
{Victor.Valdes,JoseM.Martinez}@uam.es

Abstract. Nowadays the huge amount of video material stored in multimedia repositories makes the search and retrieval of such content a very slow and usually difficult task. The existing video abstraction systems aim to ease the multimedia access problem by providing short versions of the original content which ease the search and navigation process by reducing the time spent in content browsing. There are many video abstraction system variations providing different kinds of output (video skims, keyframe-based summaries, etc). This paper presents a unified and generic video abstraction architecture which, tyring to synthesize existing apporaches from the literature, aims to characterize the stages required for a generic abstraction process as well as the definition of the theoretical aspects and requirements for the modelling of video abstraction systems which is the first step before building abstraction systems with specific characteristics.

1 Introduction

Nowadays video abstraction techniques are becoming a need in order to deal with the huge and increasing amount of audiovisual content present in home or networked repositories. The access to such amount of multimedia content is an increasingly difficult task usually implying a video preview process by the user which requires spending a large amount of time visualizing videos, as well as the bandwith required for downloading them. Those difficulties could be easily reduced with the application of video abstraction tecnhiques providing short but significative representations of the original content which could be used as previews, therefore reducing the amount of bandwidth and time required in the search and browsing processes. There exist many different approaches for the generation of video abstracts which involve different levels of complexity and abstraction algorithms. Although there exist a high heterogeneity in the different approaches, most part of the abstraction systems in the literature share conceptual stages which can be modelled in a generic video abstraction architecture. In order to be able to design such architecure, it is required first to study the characterization of the different video abstraction techniques and provide a taxonomy of them [1] attending to their external characteristics and internal characteristics. External characteristics are related to what kind of video abstract is generated

D. Duke et al. (Eds.): SAMT 2008, LNCS 5392, pp. 164–177, 2008.

and what is the behaviour of the abstraction system attending to its external interfaces and observable process characteristics: Output -the kind of abstract output, eg. video skims or keyframes-, Presentation -how the video abstract is presented-, Size -length or size of the video abstract-, Performance -amount of process required in the abstraction process- and Generation Delay -the amount of time required by the abstraction algorithm in order to start the abstract output-. Some of the different external characteristics associated to an abstraction mechanism (output size, performance or generation delay) are closely related and it is usual to find groups of abstraction methodologies sharing the same set of external characteristics. On the other side, there is a close relationship between those external characteristics and the internal methodologies applied in the abstraction process. For example, local or global applicability of the analysis and selection techniques is a very relevant factor which has a direct impact in the external observable features of an abstraction system. Therefore, internal characteristic like Basic Unit (BU) type -the smallest considered processing unit, e.g. frame, shot, sequence, etc.-, Analysis algorithms- application of analysis techniques on single basic units or between BUs-, and Rating & Selection methods - analysis and selection based on single BUs or BU comparison based rating & selection techniques-, provide also clues for designing a video abstraction architecture. Therefore, there is a close relationship between the internal characteristics of the abstract generation method and its external characteristics (efficiency, delay, etc.). The set of available abstraction techniques, considering them from an external point of view, than can be applied rely on specific internal operation modes and it is not possible to combine every abstraction methodology with any of the possible internal operation modes.

The main focus of the work presented in this paper is the definition of a common framework enabling the application and study of those techniques which will be presented in the following sections. Although there exist works in the literature which depict and classify many of the existing abstraction approaches [2,3] attending to different criteria and algorithm characteristics there are few works dealing with generic video abstraction architectures. [4] presents a conceptual framework for video summarization in which different categories of video summarization and video summaries are considered although the approach does not deal with specific abstraction stages definition. [5] presents a video abstraction system in which 3 stages are roughly defined: video segmentation and analysis, clip selection and clip assembly. In addition there exist a many specific approaches such as [6] which depicts a framework for sports video summarization (with its specific application to a soccer video scenario although several concepts could be extrapolated to generic event or highlight oriented video abstraction) or many other [7,8,9] where specific approaches are presented without trying to model the abstraction process in generic terms.

The work has been carried out looking at various existing video abstraction methods and systems from the literature (due to space constraints the list of references has been constrained to some of them) and synthesising their approaches together to generalise them into a unified model. Also due to space constraints we have not included in this paper a complete analysis of existing systems and their matching to the proposed model, but section 3.4 includes such exercise with the references in this paper.

The paper is structured as follows: section 2 depicts the proposed architecture for generic video abstraction systems modeling based on the functional division of the abstraction process on several stages. Section 3 demonstrates the applicability of the proposed architecture for modeling different abstraction systems and examples. Finally section 4 presents the conclusions of the work.

2   Architectural Models for Video Abstraction

The aim of the proposed architecture for video abstraction is to provide a modular, as simpler as possible, solution which could allow to fit inside most part of the (current and future) existing video abstraction approaches; at the same time, the different steps needed to complete a generic abstraction process will be identified. The division of the problem in independent modules will allow to further separately study the different steps involved in the abstraction process. The proposed architecture will depict the abstraction process as a chain of independent stages through which the video basic units (-BUs- as defined in section 1) "travel through" while being analyzed, compared, and selected or discarded in order to compose the video abstract. Although most practical implementations of abstraction algorithms will usually not follow a stage approach, this conceptual division is useful as theoretical framework for the study of the abstraction process mechanisms and the capabilities and performance of different abstraction models. The different abstraction stages will be considered as independent processing stages, but it will be permitted to share information and metadata about the summary generation and processed BUs between those stages. Such information will be useful for guiding the processing mechanisms of the different stages or for allowing the modeling of scoring mechanisms in which, for example, information about already selected video units could be relevant.

The proposed generic abstraction model will ease the generic study of abstraction mechanisms and the restrictions required for building systems with specific external characteristics (as defined in section 1).

2.1   Simplified Functional Architecture

As a starting point, any abstraction process can be considered as a black box where an original video is processed to produce a video summary as it is shown in figure 1 (a). Any possible abstraction algorithm may be described in this way, as it is just defined as a process which receives an original video input and outputs a video abstract. Nevertheless this model has not utility for abstraction mechanisms analysis and a more detailed model is required.

The first significative characterization of the abstraction architecture is shown in figure 1 (b) where the abstraction process is divided in two stages: 'analysis' and 'generation'. The analysis stage will be in charge of extracting relevant

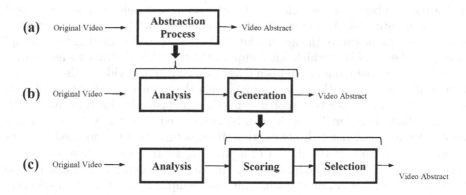

Fig. 1. Abstraction Stages Decomposition

features from the original video which could be later taken into consideration for producing the final video abstract in the 'generation' stage. This division is quite straightforward as the 'analysis' stage will include all the operations performed within any algorithm aimed to extract information from the original content with independence of the further usage of such information. Therefore all the manipulation and application of the extracted information in the abstraction process will be performed in the 'generation' stage, including BU comparison (with independence of the comparison criterion), ranking and selection.

The 'generation' stage is, in many cases, the most complex one in an abstraction algorithm and for this reason a further conceptual division will be considered as it is shown in 1 (c) where the 'generation' stage is divided in 'rating' and 'selection' stages. The rating stage will be in charge of providing a score or rank for each original video BU without carrying out any selection of BUs to be included in the output abstract. On the other hand the selection step will perform the decision about which of the incoming BUs must be written in the video abstract (and which discarded) based on the rating obtained in the previous step. The complexity balance between those two stages will depend on the specific abstraction algorithm being modeled. It is possible to find abstraction mechanisms suitable to be modeled with a very simple scoring mechanism with a subsequent application of complex selection algorithms (e.g. selection mechanisms dealing with the abstract size, continuity, score, etc. [10]) while, in other cases, the selection algorithm can be reduced to a mechanism as simple as including in the abstract those BUs rated with a value over a predefined threshold [11].

It is straightforward to demonstrate that any abstraction technique fits inside the proposed functional model just by encapsulating all the abstraction algorithm complexity in the 'rating' stage and forcing it, for example, to output '1' in case a BU should be included in the final abstract and '0' otherwise. The selection module will therefore be reduced to a process which will drop those BUs rated as '0' and will write in the output abstract the BUs with score='1'. The application of this approach makes it possible to fit any abstraction modality inside the analysis-rating-selection model, but it would be usually possible to find a balance between the different stages making the 'selection' stage to deal with the quantitative characteristics of the video abstract (e.g., size, continuity), while trying to maximize the accumulated score value (calculated in the 'rating' stage) for the set of BUs which will compose the resulting abstract without other considerations about the mechanism employed to rate the video BUs (this case is specially suitable for those systems in which the score associated to a BU will not be affected by other BUs scores or by their inclusion in the video abstract).

The defined functional modules can be considered as a minimal set of stages needed for a generic enough abstraction architecture: there is no need for considering all of them in the design of a working abstraction system, being the 'selection' stage the only mandatory one (a minimal abstraction system can be built with a single selection stage in which subsampling [12] or random selection of BUs could performed) although most of the existing abstraction approaches can be modeled with this 'selection'- 'analysis'-'scoring' stages approach.

2.2   Generic Video Abstraction Architecture

Figure 2 depicts a generic abstraction architecture following the analysis-scoring-selection functional modules architecture. The data flow between the different stages is shown, as well as the repositories/databases aimed for metadata storage which will be considered as information repositories with independence of the mechanisms applied for its implementation (direct memory access, disk storage, databases, etc.).

The 'reading', 'selection' and 'writing' steps are mandatory in any abstraction architecture: a way to read and write video (disk storage, memory buffers, video streaming, etc.) and a mechanism to output part of the original video content in order to generate the video abstract must be included in any abstraction system (although they don't conform characteristic features of any considered abstraction approach). The rest of the stages -'analysis', 'scoring' and 'presentation'- are not mandatory: for example, in the case of a video abstraction system based on a direct uniform or random subsampling of the original video, there is no

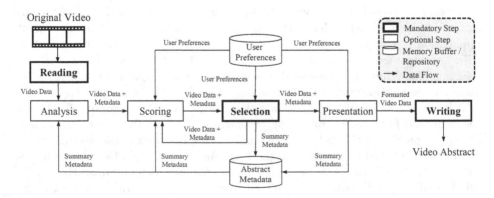

Fig. 2. Generic Abstraction Architecture

need to analyze or rate the original content, being possible to implement the subsampling mechanism directly in the 'selection' stage. A 'presentation' step has been included in the genericarchitecture for providing coverage to those abstraction approaches in which video editing or formatting is needed, but in this paper it will be considered as a secondary stage when analyzing abstraction mechanisms.

The abstraction process will be considered as the flow of BUs through the different stages included in the depicted architecture until all the available BUs have been processed (that is, selected to be included in the output abstract or discarded) and therefore the abstraction process is finished. All the processing stages (analysis, scoring, selection and presentation) will receive BUs which can be accumulated, processed (analyzed, annotated, rated, combined, etc.), and then redirected to other stage, discarded or selected to be included in the output abstract. Any stage can produce metadata (e.g. low-level video analysis results, mid-level feature extraction results, set of included/discarded BUs, BU classification in different content categories, etc.) that can be appended to the BUs and/or stored in the 'Abstract Metadata' repository, so that this information may be used in other stages, such as the 'analysis' or 'scoring' ones, allowing a feedback information mechanism.

In many cases the BUs will be constituted only by the original video content (frames, GoPs, shots,...) but, as result of the different processing stages, metadata can be appended to the original content. Usually, once a video BU is analyzed and rated, it is either selected (being its associated video content written to the output summary) or discarded (deleting its associated original video content), but there are specific abstraction algorithms (e.g., see [13]) in which the selection or discard of a specific BU as part of the output abstract yields to the recalculation of other previously calculated BUs scores. Figure 2 shows a video and metadata feedback flow from the selection stage to the scoring stage in order to allow the modeling of such systems. As will be shown in the following sections this kind of approach enables the development of different abstraction modalities (see section 3)but it hasa negative impact on the efficiency of thesystem.

The 'User Preferences' repository has been included in the architecture in order to make available user preferences or system configurations for those cases in which this information can be applied for the generation of customized video abstracts. The user can define preferences about different abstract characteristics such as its length (or size) to be considered as an additional constraint in the selection stage, its modality (e.g., story-board, skim, video-poster) or media format for the output abstract to be considered by the presentation stage, or what category of content (e.g., economy, sports, wheater forecast) the user is interested in to be used in the selection stage as a filtering constraint.

3   Abstraction Systems Modeling

A generic stage-based functional architecture for video abstraction systems has been depicted in the previous section. Such architecture is based in the consideration of the abstraction process as the flow of independent input BUs through the defined stages of a specific abstraction system until the complete set of original video BUs has leaved the system. The set of defined abstraction stages, what kind of BU is considered in the system, the way in which each BU is processed in each stage, how much time it is needed for this process, how many BUs each stage can keep in order to complete its processing and other considerations will vary depending on the specific abstraction approach and will determine the external and internal abstraction system characteristics (previously defined in section 1) and viceversa. In this section a some common abstraction approaches will be studied, depicting the different stages and characteristics needed for its implementation following the proposed architecture and abstraction approach.

The different approaches will be grouped in non-iterative systems, that is systems where the BUs are processed at the most one time per stage, and iterative systems, where the BUs can be resent to the 'scoring' stage after being processed by the 'selection' stage.

3.1   Non-iterative Video Abstraction Systems

Figure 3 (a) depicts the most simple abstraction architecture possible. In this case neither 'analysis' or 'scoring' stages nor metadata interchange mechanisms are needed for the abstraction process as it is based in a subsampling mechanism. All the needed algorithms are included in a 'selection' stage in charge of selecting 1 out of every n BUs (e.g. just by direct subsampling or random selection) guided by the 'User Preferences' which, in this cases, will be limited to the selection of the output abstract rate $\frac{1}{n}$. This kind of systems are able to generate progressive, bounded size abstracts with negligible delay and linear performance as no analysis over the original content is needed and only a simple selection process is employed. An example of this architecture can be found in [12].

Using the same architecture and just considering different kind of input and output BUs, keyframe or video skim outputs can be generated and even formatted in different ways if a 'presentation' stage (see section 2) is appended to the system; this consideration can be generalized for any of the abstraction systems that will be described next.

Figure 3 (b) shows a possible architecture for abstraction systems based on content analysis in which the rating of the BUs depends only on the original content. In this case the 'analysis' and 'scoring' stages are included in the system. The 'analysis' stage is in charge of extracting features from the original video which can be later used in the 'scoring' stage to rate the incoming BUs. Based on the previous BU rate, the 'selection' stage selects the BUs to be included in the output abstract taking into account the 'User Preferences' which could specify, for example, the desired abstract length. Such 'User Preferences' may also be used in the 'Scoring' stage for personalization purposes. It should be noted that this model does not make use of the 'Abstract Metadata' database which enables a mechanism for feedback between the different stages and therefore only rating mechanisms which are independent of the under generation abstract (that is, the

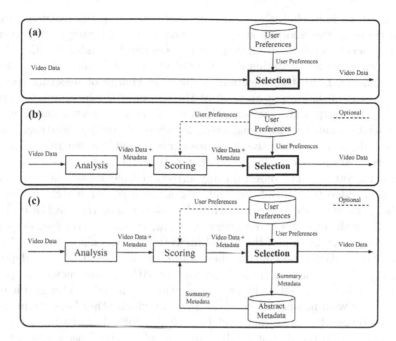

Fig. 3. Non-iterative Abstraction Architecture Examples

score or rate for each incoming BU does not depend on the previous inclusion of other BUs) can be modeled.

Adaptive subsampling abstraction systems (e.g., [14] or [15]) and Relevance curve-based abstraction systems (e.g. [16], [17]) can be modeled with the architecture depicted in 3 (b). Adaptative subsampling abstraction systems usually perform a selection of a subset of the original BUs based on a varying sampling rate which, in this case, does not rely in a simple control of the output abstract size but in a previous analysis of the original content. The system described in [14] outputs keyframe abstracts performing non-uniform sampling based on motion activity and it can be modeled with the proposed architecture including an initial 'analysis' stage, that extracts the motion activity of the incoming BUs (the original video is divided in shots as input BUs), followed by the 'scoring' stage in charge of rating each incoming BU with the acumulated extracted motion activity for each frame composing the BU (no 'User Preferences' are considered in this case). Finally the 'selection' stage performs a selection of n frames from each shot based on its accumulated motion activity value (for example by an uniform division of the accumulated motion value selecting the frames corresponding to the division positions). The model can be generalized for any kind of adaptive sampling abstraction mechanism by considering different extracted features and rating mechanisms. Relevance curve-based abstraction systems calculate a score value for each BU based on different criteria such as motion activity, face detection, audio energy, ... and generate a relevance curve which is later used for the selection of the video abstract BUs. In the systems described in [16], [17] the

'Analysis' stage is in charge of analyzing and extracting the features from the original content, the 'scoring' stage is in charge of combining such features in order to generate a relevance curve (this process can be guided by 'User Preferences' for example for weighting the different extracted features in the relevance curve calculation) and the 'selection' stage is in charge of selecting the set of BUs to be part of the output abstract as the subset of the incoming BUs is constrained by the output abstract size which maximizes the accumulated relevance (other variables can be also considered, for example output abstract continuity). It should be noted that the relevance curve model is a generalization which includes several approaches for video abstraction due to the possibility of generating a relevance curve based on any criterion such as salient fragments of the original video (e.g. people present, a goal in a soccer match, motion in a surveillance camera, etc.) or considering the relevance as the level of fulfillment of each BU with respect a given query (textual, visual, by concept, etc.) which enables the generalization of the model for retrieval systems.

Those abstractions mechanism where the scoring of each BU is independent from other BUs or have a limited number of BUs dependencies are particularly suitable for progressive abstract generation as the BUs enter and leave the defined stages with independence of the state in which other BUs are in the system. Nevertheless the presented model is suitable for other abstraction systems in which this condition is not fulfilled such as clustering based systems [7,18]. In a typical clustering algorithms a first stage of data preprocessing over the original data is carried out in order to extract features as basis for a subsequent clustering process. Those two stages are clearly identified with the 'analysis' -data preprocessing- and 'scoring' stage in which the input data are clustered (the clustering process can be considered as a grouping process in which low-level BUs -frames, shots,...- are grouped in higher level BUs -clusters-) and scored (for example considering its distance to the cluster centroid or a representativeness value). The further 'selection' stage will be in charge of selecting the final set of BUs from the calculated clusters, for example selecting the closest BUs to each cluster centroid [19]. In this case the whole set of original video BUs is needed in order to complete the clustering process in the 'scoring' stage and hence the clustering approach is not suitable for progressive abstract generation as no BU leaves the 'scoring' stage until all the original content has been processed.

Figure 3 (c) shows a variation of the architecture in which the 'Abstract Metadata' database has been enabled. This database is a representation of any possible information feedback mechanism between the 'selection' and 'scoring' stages aimed to provide information about the generated abstract needed by certain abstraction approaches. For example, sufficient content change (as called in [3]) abstraction approaches [20] which include BUs in the output abstract only if its visual content differs significantly from previously extracted keyframes need a mechanism to access to previously stored BUs' visual information. In this case the 'analysis' stage extracts visual features from the original content such as color histograms [11] or the MPEG-7 Color Layout descriptions [21] which are linked to the incoming BUs. In the 'scoring' stage the BUs are rated attending

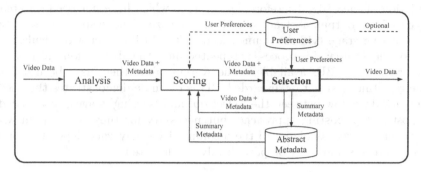

Fig. 4. Iterative Abstraction architecture

to its similarity to previous BUs already included in the output abstract, whose extracted visual features are available in the 'Abstract Metadata' repository. In the 'selection' stage, and based on the previously obtained score, the incoming BUs are discarded or included in the output abstract (and in this case its associated extracted visual information is stored in the 'Abstract Metadata' repositoty for its usage in the following BUs ratings.

3.2 Iterative Video Abstraction Systems

The models shown in figure 3 are valid for abstraction systems where the incoming BUs are processed only once by each defined stage and permit to model most of the existing abstraction approaches. Nevertheless there are cases of abstraction systems based on iterative processing [22][13] which require an architecture as the one shown in figure 4 where there is an additional video data (BUs) and metadata flow from the 'selection' to the 'scoring' stage, allowing the iterative scoring of the BUs. The maximum frame coverage abstraction approach[13][23] tries to generate a video abstract selecting a set of BUs as most representative as possible of the original content accordling to a specific criterion (usually visual similarity). For example in [13] the coverage of each original frame is calculated as the set of all frames in the original video which are considered as similar to the given one and iteratively the most representative frame (the frame which a higher number of similar frames) is selected to be included in the output abstract, removing the set of similar frames from the available frames for selection. This scheme can be easily modelled with the proposed architecture as a frame-BU system with an initial 'analysis' stage where visual features are calculated (e.g., color histograms, color layout), a 'scoring' stage where the coverage value (i.e. the set of similar BUs to each given one) is calculated and a 'selection' stage where the BUs for the video abstract are iteratively selected as those with the highest coverage values. Each time a BU is selected there is a set of the original BUs -those similar to the selected one- which are considered as already 'represented' in the output abstract and must be eliminated from the set of BUs available to be included in the output and not considered anymore in the coverage value calculation of each BU. Therefore, all the BUs which have not been selected to be part of the output abstract nor eliminated from the system are sent back to the scoring stage (via the defined feedback data flow) for a recalculation of their coverage value; this process is repeated until the abstraction finishes when there are no more BUs available in the system (all of them have been included in the resulting abstract or discarded). This example demonstrates the need of a feedback data flow between the 'selection' and 'scoring' stages, not needed in the most of the existing approaches but necessary for those systems in which the original scores associated to the incoming BUs may vary depending on the selection process and must be periodically recalculated.

3.3   Considerations with Respect to the Presentation Stage

In order to complete the architecture proposed in section 2(shown in figure 2) the 'presentation' stage should be considered in the depicted models but it has been found that it is rarely present in the studied abstraction systems and has neither impact in the BUs selection methods nor in the overall system efficiency. Therefore, in this work it has been considered as a secondary element which can be appended to the system but has no relevance in terms of systems characterization. In the same way, the video and metadata flow between the 'presentation' and 'analysis' stages have been added in 2 (including 'User Preferences' for the 'presentation stage') for completeness reasons and enabling, for example, the possibility of modeling adaptive analysis mechanisms or different presentation approaches guided by the characteristics of the generated abstracts (size, score, continuity, etc.).

3.4   Abstraction Systems Architecture Summary

In this section the set of referenced abstraction systems are summarized considering the different categories previously depicted attending to its complexity, which corresponds with the set of modules and dataflow from the complete generic architecture depicted in section 2.2 that are needed for building such system. The computational complexity of every abstraction system will directly depend on each abstraction module internal complexity as well as the architectural category in which it is included.

Other abstraction systems implementable with different combinations of abstraction modules and dataflows may exist in the literature: the different presented combinations have been selected because they are considered as representative models. The enumerated systems can, in several cases, be modelled with variations in the balance between the tasks corresponding to each abstraction module but, in any case, if such modelling is carried out taking into account the considerations enumerated in section 2 the systems should be included in the same architectural category.

Table 1 summarizes the different architectural categories divided in Iterative -I- or Not Iterative -NI- and depicting the set of abstraction modules which are

Table 1. Abstraction System Architectural Classification

ARCHITECTURAL CATEGORY	REFERENCES
[NI,Sl]	[12]
[NI,A,Sc,Sel]	[7,14,15,16,17,18,19]
[NI,A,Sc,Sel,MF]	[11,21,20]
[I, A, Sc, Sel, MF]	[13,22,23]

included: Analysis -A-, Scoring -Sc-, Selection -Sel- and the inclusion of a Meta-data Feedback -MF- dataflow. It is not surprising to find a lack of references corresponding to the [NI,Sl] category due to its simplicity (e.g. simple subsampling or selection of the beginning of the original video) while a number different approaches are found in the rest of the categories.

4   Conclusions

An architecture which allows the development of generic abstraction systems as a sequential processing of the original video BUs has been proposed. This architecture starts from isolating the different possible stages involved in the video abstract generation process, considering each stage as BU processing modules in charge of analyzing, adding information and redirecting the incoming BUs. This separation between the different abstract generation stages will allow the generic study of the abstraction algorithms by dividing the different approaches and studying each part independently while, at the same time, enabling the development of generic interchangeable modules for the analysis, scoring, selection and presentation algorithms to be combined in different ways for experimentation. The study and characterization of each independent step in the abstraction chain will allow future understanding of complex systems starting from the independent study of each stage. Once we have a good understanding of video abstraction processes and have a standardised exchange established, the following scenario can be a reality: system A, developed by a video processing group, has a strong Analysis module - if input/output of the module is in standardised format, then system B, developed by a media producer, could borrow that module for its analysis and use system B's fancy presenation module to output a better abstraction.

The proposed approach will ease the task of analyzing the performance and internal/external characteristics (see section 1) of any proposed system in an unified framework which could be applied for subsequent systems comparison and characteristics specification as well as the classification of the different abstraction approaches attending to their architectural requirements. Additionally, the proposed architecture has allowed to define a set of elemental abstraction models, which are suitable for building almost any of the most spread abstraction approaches found in the literature. The proposed model can be used for self-reflection when evaluating and comparing our methods and systems with the ones from other research groups (e.g., what my group has been focusing so far is actually the 'Scoring' stage, while that other group's system has a strength in the 'Presentation' stage).

Acknowledgements

Work supported by the European Commission (IST-FP6-027685 - Mesh), Spanish Government (TEC2007-65400 - SemanticVideo) and Comunidad de Madrid (S-0505/TIC-0223 - ProMultiDis-CM), by the Consejería de Educación of the Comunidad de Madrid and by the European Social Fund.

The authors would like to thank the anonymous reviewers, specially the meta-reviewer, for their valuable comments and forward-looking vision.

References

1. Valdés, V., Martínez, J.M.: A Taxonomy for Video Abstraction Techniques. Technical report (available upon request)
2. Li, Y., Lee, S.-H., Yeh, C.-H., Cuo, C.-C.J.: Techniques for Movie Content Analysis and Skimming. IEEE Signal Processing Magazine 23(2), 79–89 (2006)
3. Troung, B.T., Venkatesh, S.: Video Abstraction: A Systematic Review and Classification. ACM Transactions on Multimedia Computing, Communications and Applications 3(1), 1–37 (2007)
4. Money, A.G., Agius, H.: Video Summarization: A Conceptual Framework and Survey of the State of the Art. Journal of Visual Communication and Image Representatiom 19(2), 121–143 (2008)
5. Lienhart, R., Pfeiffer, S., Effelsberg, W.: Video Abstracting. Communications of the ACM 40(12), 54–62 (1997)
6. Li, B., Pan, H., Sezan, I.: A General Framework for Sports Video Summarization with its Application to Soccer. In: Proceedings of IEEE International Conference on Acoustics, Speech and Signal Processing 2003, ICASSP 2003, vol. 3, pp. 169–172 (2003)
7. Hanjalic, A., Zhang, H.J.: An Integrated Scheme for Automated Video Abstraction Based on Unsupervised Cluster-Validity Analysis. IEEE Transactions on Circuits and Systems for Video Technology 9(8), 1280–1289 (1999)
8. Ma, Y.-F., Zhang, H.-J., Li, M.: A User Attention Model for Video Summarization. In: Proceedings of ACM Multimedia 2002, pp. 533–542 (2002)
9. Ciocca, G., Schettini, R.: An Innovative Algorithm for Key Frame Extraction in Video Summarization. Journal of Real-Time Image Processing 1(1), 69–88 (2006)
10. Fayzullim, M., Subrahmanian, V.S., Picariello, A., Sapino, M.L.: The CPR Model for Summarizing Video. In: Proceedings of the 1st ACM International Workshop on Multimedia Databases, pp. 2–9 (2003)
11. Valdés, V., Martínez, J.M.: On-line Video Skimming Based on Histogram similarity. In: Proceedings of ACM Multimedia 2007 - International Workshop on TRECVID Video Summarization, pp. 94–98 (2007)
12. Hauptmann, A.G., Christel, M.G., Lin, W.-H., Maher, B., Yang, J., Baron, R.V., Xiang, G.: Clever Clustering vs. Simple Speed-Up for Summarizing BBC Rushes. In: Proceedings of ACM Multimedia 2007 - International Workshop on TRECVID Video Summarization, pp. 94–98 (2007)

13. Chang, H.S., Sull, S., Lee, S.U.: Efficient Video Indexing Scheme for Content-Based Retrieval. IEEE Transaction on Circuits and Systems for Video Technology 9(8), 1269–1279 (1999)
14. Divakaran, A., Radhakrishnan, R., Peker, K.A.: Motion Activity-Based Extraction of Key-Frames from Video Shots. In: Proceedings of IEEE International Conference on Image Processing 2002, vol. 1, pp. 932–935 (2002)
15. Nam, J., Tewfik, A.H.: Video Abstract of Video. In: Proceedings of IEEE third Workshop on Multimedia Signal Processing, pp. 117–122 (1999)
16. Valdés, V., Martínez, J.M.: Post-processing techniques for on-line adaptive video summarization based on relevance curves. In: Falcidieno, B., Spagnuolo, M., Avrithis, Y., Kompatsiaris, I., Buitelaar, P. (eds.) SAMT 2007. LNCS, vol. 4816, pp. 144–157. Springer, Heidelberg (2007)
17. Albanese, M., Fayzullin, M., Picariello, A., Subrahmanian: The priority curve algorithm for video summarization. Information Systems 31(7), 679–695 (2006)
18. Ratakonda, K., Sezan, M.I., Crinon, R.J.: Hierarchical Video Summarization. In: Proceedings of Visual Communications and Image Processing 1999, SPIE, vol. 3653, pp. 1531–1541 (1999)
19. Yu, X.-D., Wang, L., Tian, Q., Xue, P.: Multi-Level Video Representation with Application to Keyframe Extraction. In: Proceedings of 10th International Multimedia Modelling Conference 2004, pp. 117–123 (2004)
20. Kim, C., Hwang, J.-N.: Object-Based video abstraction for video surveillance systems. IEEE Transactions on Circuits and Systems for Video Technology 12(12), 1128–1138 (2002)
21. Valdés, V., Martínez, J.M.: On-Line Video Summarization Based on Signature-Based Junk and Redundancy Filtering. In: Proceedings of 9th International Workshop on Image Analysis for Multimedia Interactive Services, WIAMIS 2008, pp. 88–91 (2008)
22. Hun-Cheol, L., Seong-Dae, K.: Iterative Keyframe Selection in the Rate-Constraint Environment. Signal Processing: Image Communication 18(1), 1–15 (2003)
23. Yahiaoui, I., Merialdo, B., Huet, B.: Automatic Video Summarization. In: Proceedings of MMCBIR 2001, Multimedia Content Based Indexing and Retrieval, Rocquencourt, France (September 2001)

Author Index